Macgregor's
Excu
to th
GEOL
ARRAN

Third Edition
Revised and edited by

J. G. MacDONALD and A. HERRIOT

with contributions by
B. C. KING

published by the

GEOLOGICAL SOCIETY OF GLASGOW, 1983

THE UNIVERSITY, GLASGOW G12 8QQ

Printed by
McCORQUODALE (SCOTLAND) LTD
GLASGOW

First edition 1965
Reprinted 1968
Second edition 1972
Third edition (Revised) 1983

ISBN 0 902892 07 X

PLATE I. The Northern Granite Mountains. (*For explanation, see page 8*)

FOREWORD

Arran, that lovely island which dominates the broad estuary of the Firth of Clyde, has long been a favourite resort of the holidaymaker, drawn to it by its diversity of scenery and the grandeur of its northern peaks. To the geologist it has been for many years a place of pilgrimage, offering as it does within a comparatively small area a wide variety of rock formations and many striking, and indeed classic, illustrations of geological succession and structure.

The geology of the island has gathered round it over the years a considerable literature. This excursion guide has been designed to serve as an introduction to the subject, and it is hoped that it will prove of use to students of geology as well as of interest to those who find a perennial attraction in the scenic charm of the island.

The death of Dr. Murray Macgregor occurred in January 1966, shortly after the publication of the first edition of this guide. A short biography and list of other publications of this eminent geologist can be found in the "Proceedings of the Geological Society of Glasgow", session 108 (1967), pp. 17–24.

FOREWORD TO THE 3rd EDITION

Arran continues to hold a fascination for geologists, both amateur and professional, but over the years since this guide was first published several important developments have taken place in basic understanding of geological principals and in our knowledge of the Arran rocks. Also the activities of the Forestry Commission have altered the physical access in some areas.

In this first major revision of Macgregor's Guide the opportunity has been taken to incorporate a number of changes which reflect these developments; at the same time measurements have been modified to take account of metrication and many of the figures have been re-drawn. Throughout, however, an effort has been made to preserve as much as possible of the original text so that the third edition can still be substantially thought of as the work of Dr. Macgregor.

CONTENTS

ILLUSTRATIONS

PLATES

1 INTRODUCTION

The Guide has been planned to provide an introduction to the geology of Arran mainly by means of a series of descriptive itineraries for which brief notes on selected localities have been prepared. Two criteria have been kept in view in the selection of excursion routes and localities alike, first, their importance in furnishing illustrations of the broad features of the geology of the island and, second, their accessibility. In order to link together the various excursions planned a short account of the geological history of Arran has also been included and, to cater especially for those who have only a limited time available to spend on the island, there are road logs which contain some localities in addition to those mentioned in the excursions.

Arran has a perennial fascination for the geologist. Since the days of James Hutton (1726–1797) it has drawn students and research workers from many lands, attracted to the island by its diversity of scenery, by its wide range of formations and by the profusion and variety of its igneous rocks. Thirty-two kilometres in length and from thirteen to seventeen kilometres in breadth, it contains within the small compass of about four hundred and thirty-five square kilometres sedimentary rocks belonging to nine different major geological divisions ranging in age from the unconsolidated deposits of Recent and Glacial times to the ancient crystalline schists to which the name Dalradian has been given. The table set out on page 21 shows the order of succession of these rocks and also indicates how immensely prolonged is the period of geological time they cover. In addition to rocks of sedimentary origin, nearly half the area of the island is occupied by igneous rocks of various ages and of very diverse nature.

It must be emphasised that the Guide is merely an introduction to the geology of an extremely complex district, a district which despite all the research work that has been done still holds many unsolved problems. Its main purposes are to assist the beginner in classifying the various types of rock entering into the structure of the island and in studying their mode of occurrence, distribution and relationships. The notes attached to individual localities on the excursion routes must necessarily be brief; the student should make his own observations and notes and will certainly find much to add to the descriptions given here. Within these limits, however, it is hoped that the excursion maps and locality notes will serve both as an introduction to the field geology and as a stimulus to further independent work. Some of the less common technical terms used in these notes are explained where they occur: for the meaning of any terms which the reader does not understand reference should be made to some general geological text-book or to a dictionary of geological terms (see, for example, Whitten and Brooks 1972 in the list of references).

Summary of Physical Features: Much has been written on the subject of the variety and charm of the Arran landscape. From a geological point of view a description of its physical features (or physiography) implies some enquiry into their origin and interrelationships. It becomes then part of the later geological history of the island, a subject briefly dealt with in a later section of this book. Here all that is attempted is a short account of its main topographic features in relation to their present geological setting. It should be emphasised that the configuration of Arran is due essentially to two factors; (1) the vast time interval throughout which it has been part of a land surface and subjected to the various agencies of denudation—weathering, frost-action and erosion by running water and ice; and (2) the varying resistance offered to these agencies by rocks of widely different origin, composition and degrees of hardness. Scotland has been a land surface ever since the floor of the late-Cretaceous (Chalk) sea was elevated above sea level. To the long interval of time during which its physiography was slowly evolved the name of the Cainozoic Era has been given. This era is estimated to have lasted about 65 million years and it was not until near its close that Arran became

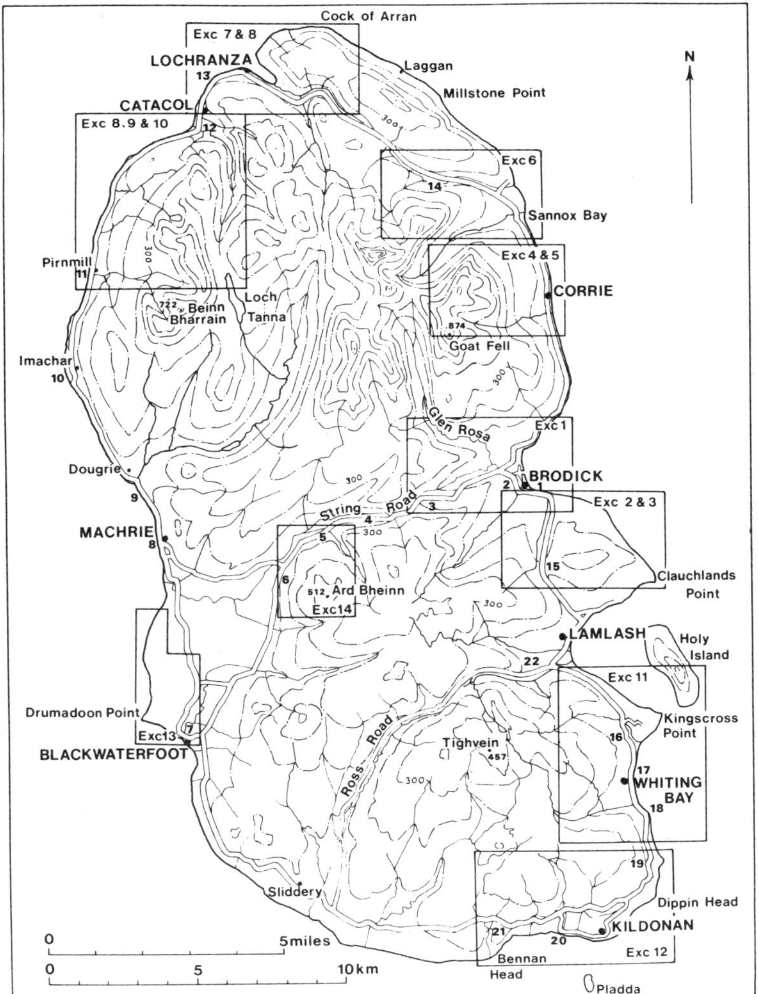

FIG. 1. Simplified physiographic map of Arran showing areas covered by the excursion itineraries. Road log localities indicated by numbers. Contour interval 100 metres.

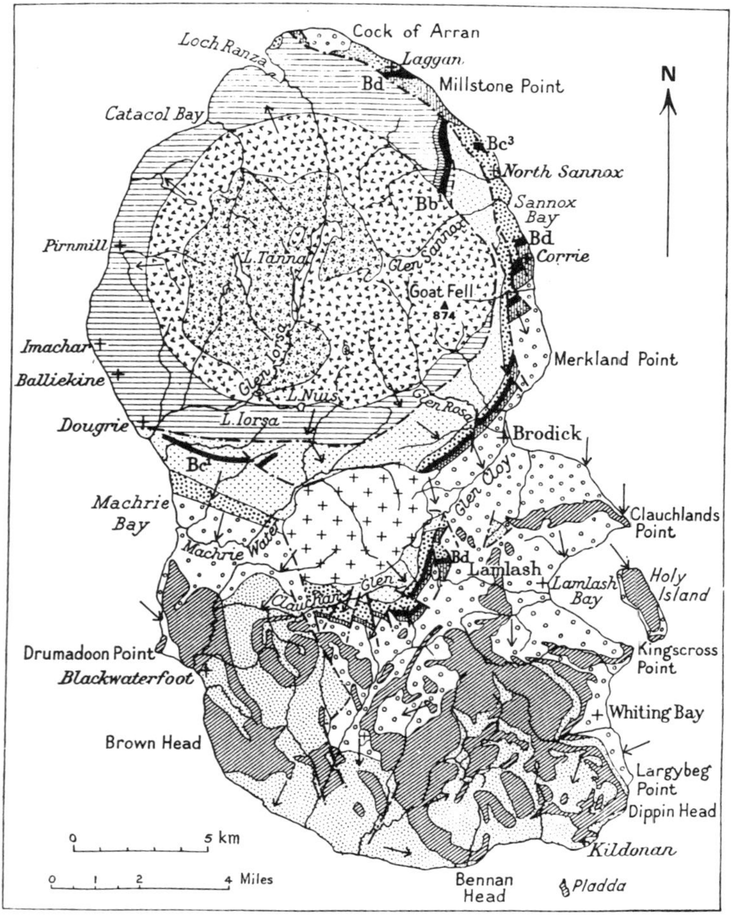

FIG. 2. Simplified geological map of Arran; based on the Geological Survey one-inch map.

an island. It was even later still that Britain was separated from mainland Europe.

A comparison of Figures 1 and 2 shows that Arran is divided into two nearly equal portions, presenting strongly contrasting types of topography and geological structure, by a line drawn from Brodick to Dougrie. The northern half is rugged and mountainous, with an essentially highland type of scenery; the southern half has a much more subdued relief and presents an undulating, plateau-like surface.

In the northern half of the island the dominant structural unit is the nearly circular mass of the Northern Granite, entirely interior and covering an area of some 145 square kilometres. It is out of this granite that the high peaks, the serrated ridges and the massive buttresses of the northern mountains have been carved. The highest ground occurs east of Glen Iorsa, where the mountains fall into four main groups centred around Cir Mhòr and culminating in Goatfell at an elevation of 874 metres. The heights of the chief summits of this group and the heights of some of the principal mountain passes or cols are shown in Figure 3. West of Glen Iorsa the most striking scenic feature of the granite massif is the long, more or less continuous, ridge of high ground lying west and northwest of Loch Tanna and overlooking Kilbrannan Sound. This ridge includes, from south to north, Beinn Bharrain with its twin summits of 715 and 722 metres; Beinn Bhreac, 712 metres; Meall Bhiorach, 549 metres, rising above picturesque Coire an Lochan; and Meall nan Damh, 570 metres.

INDEX TO FIGURE 2

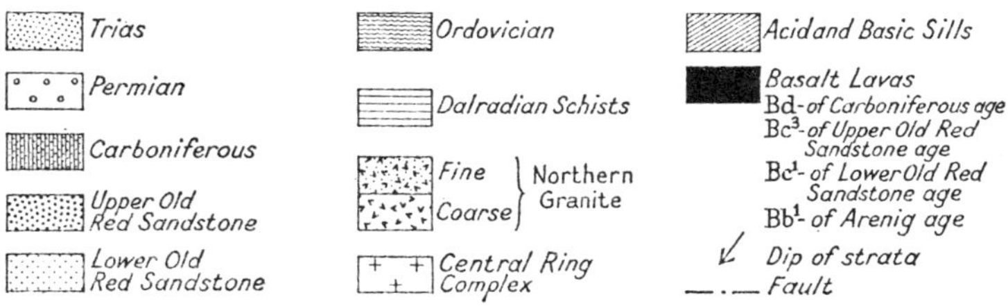

The Northern Granite is surrounded by a ring of older rocks into which it has been intruded from depth as a huge more or less vertical dome-shaped mass. Immediately in contact with it around its margin, except for a distance of about 5 km on its eastern side (from a little south of the Corrie Burn or White Water to the vicinity of Cnocan Donna in Glen Sannox) are the Dalradian Schists, the oldest rocks in Arran (see page 20). These occupy the west coast from Dougrie to Loch Ranza and form the rough craggy ground north and northeast of Glen Chalmadale, up to 443 m. The Dalradian rocks are followed eastwards and outwards from the granite by the steeply-dipping Old Red Sandstone, succeeded in turn by the Carboniferous lavas and sediments well exposed at Corrie, and then by the Permian sandstones and breccias which form the coast between Sannox and Brodick. A narrow strip of Carboniferous rocks, faulted against Dalradian, stretches from near the Fallen Rocks for some 5 km to the northwest until overlain in this direction by the Permian breccias of the Cock of Arran and An Scriodan.

The southern half of the island has a very different topography and structure. Apart from an area of some 18 square km extending from the central part of the String Road southwards to the headwaters of the Clauchan Glen to which reference is made later, it consists of an undulating tableland with a maximum elevation of 457 metres attained at Tighvein 4·8 km southwest of Lamlash. It is floored by sedimentary rocks belonging to the Permian and Triassic Systems, into which have been intruded the numerous sills and dykes which form so characteristic a feature of South Arran geology. Where the more massive sills reach the coast they give rise to the striking headlands of Clauchlands Point, Kingscross Point, Dippin Head, Brown Head and Drumadoon Point. The greater part of Holy Island, 4 km long and about 800 metres wide and rising steeply to a height of 314 m is also formed of a thick sheet or sill intruded into Permian sandstones and the little island of Pladda off the south coast about 800 m long and 21 m high is in the main a sill. All these sills and dykes must of course be of later date than the strata they cut. Actually the time interval separating

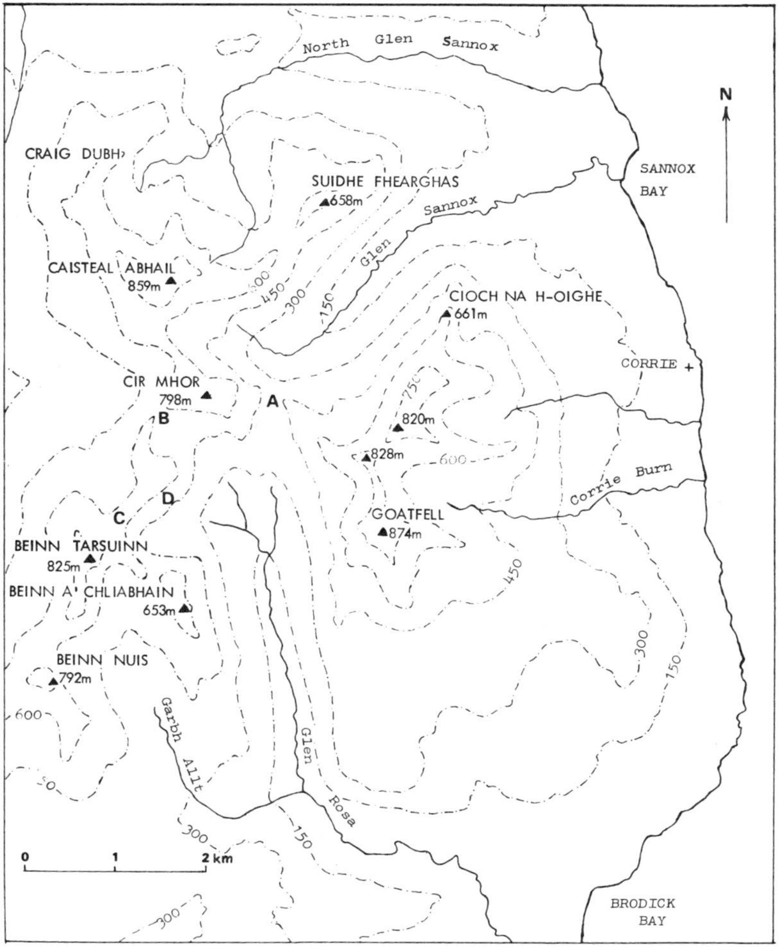

FIG. 3. The granite mountains of North Arran. Contour interval 150 m. A. The Saddle, 431 m; B. The Archer's Pass, a little over 610 m; C. unnamed pass, about 610 m; D. A'Chir ridge, about 713 m.

the deposition of the Triassic sediments from the period of intrusion of the sills and dykes covers all the later part of Mesozoic times and part of Tertiary times (see table 1, p. 21).

Reference has been made to the tract of country lying between the headwaters of Clauchan Glen and the String Road near Glenloig. Here there is an oval-shaped area some 4·8 km in diameter centred around Ard Bheinn (512 m), A'Chruach (512 m) and Beinn Bhreac (503 m) and consisting mainly of an assemblage of igneous rocks of various types and of diverse origin. This area marks the site of a deeply-dissected volcanic centre or centres of Tertiary age, and since it reproduces on a smaller scale some of the features of the great central intrusion complexes of the Inner Hebrides with their ring-shaped structures it has been named the Central Ring Complex.

The valleys of Arran fall into two categories, major and minor. The former, represented by the "through" valleys which cut across the framework of the island are:

(1) The NW–SE valley which holds the present Chalmadale and North Sannox streams, with a watershed at about the 195 m level.

(2) The three valleys traversing the Northern Granite in a general N–S direction. These are, from east to west: (*a*) the Glen Rosa hollow continued across the divide of the Saddle (431 m) into Glen Sannox; (*b*) the depression occupied by the Iorsa Water, the longest stream in the island (about 13 km) continued across the col on which Loch na Davie lies (about 360 m) into the valley of the Easan Biorach; and (*c*) the valley defined by the Catacol Burn and the 1·6 km long hollow of Loch Tanna, following a shatter belt in the granite and perhaps continued to the south-southwest by Glen Scaftigill.

(3) The WSW–ENE valley of Glen an t-Suidhe and Glen Shurig which traverses the centre of the island, with a watershed at about the 234 m level.

All these through valleys must be regarded as portions of an early drainage pattern (or patterns) long since broken up and segmented.

The minor streams take their origin for the most part on the high ground forming the central parts of the island: (i) draining the northern mountains, like the Allt Gobhlach on the west coast and the Corrie Burn on the east; (ii) radiating from the dome-shaped area of the Central Ring Complex, like Glen Craigag (a tributary of the Machrie Water) and Clauchan Glen; or (iii) rising on the slopes of the Tighvein plateau, like the Kilmory Water flowing south to the coast, the Allt Dhepin–Glenashdale Burn entering Whiting Bay and the Monamore Burn entering Lamlash Bay.

Two striking features of the topography which cannot fail to impress the visitor are: (1) the broad well-marked platform surrounding the northern mountains, the 300 metre ("Thousand-foot") Platform (p. 57); and (2) the low-level raised beach platform, with its abandoned sea cliffs and caves, which fringes so much of the Arran coast (p. 65). These represent episodes or stages, the former ancient, the latter comparatively recent, in the evolution of the existing topography.

Sources of information available: These may be grouped under two categories:

(*a*) **Maps.** The geological survey of Arran was carried out on the scale of six inches to a mile by W. Gunn between the years 1892 and 1901. Gunn's work, to the thoroughness and accuracy of which later workers have abundantly testified, was embodied in Sheets 13 and 21 of the Geological Survey one-inch map of Scotland, issued respectively in 1898 (second edition 1902) and 1901; drift editions of these sheets appeared in 1906. They were replaced, so far as Arran is concerned, by the Special Arran Sheet of 1910 (colour-printed). Later editions of this map, with additional information by G. W. Tyrrell and others, were published in 1947 and 1961 in separate solid and drift sheets. The drift editions of these show the distribution of the loose unconsolidated deposits, such as those transported by glaciers or by the meltwaters of glaciers and left behind at the close of the Glacial Period as a covering over much of the solid rock. In 1971 a new provisional 1:50,000 edition of the geological map was published but this does not incorporate any new information being a photographic enlargement of the previous "one inch" edition. At the time of writing a revised 1:50,000 edition is under preparation.

The most useful topographic map of the island is the special sheet on Arran of the Ordnance Survey 1:50,000 Map of Great Britain. A 1:40,000 map of Arran was also produced for the Orienteering Championships which were held there in 1980.

(*b*) **Memoirs and other Publications.** Two Geological Survey Memoirs that deal with Arran geology have been published. Gunn's account of the northern part of the island was included in the *Geology of North Arran, South Bute and the Cumbraes* (Explanation of Sheet 21) which appeared in 1903. Gunn died in 1902, however, before the corresponding Explanation of Sheet 13 had been completed. The comprehensive memoir on the geology of Arran by G. W. Tyrrell was published by the Geological Survey in 1928 and is still the standard and authoritative work on the subject. It incorporates much of Gunn's published and unpublished descriptions, but includes the results of the researches carried out by Tyrrell himself over many years, mainly on the complex assemblages of igneous rocks of Tertiary age. The same memoir contains an exhaustive bibliography of literature dealing with Arran geology. Any publications of later date than 1928 have been included in the list of references on pages 195–204 of this Guide.

Transport and accommodation: Most parts of the island can be reached by public transport during the summer months but from November to May bus services are less extensive. Up-to-date details of public transport may be obtained from the Tourist Office in Brodick or by contacting Arran Coaches in Brodick (tel. 0770 2121). A variety of accommodation in hotels, private houses, self-catering cottages, and caravans, is available. A list of places to stay and advance booking facilities are provided by the Isle of Arran Tourist Association, Tourist Information Centre, The Pier, Brodick, Isle of Arran (tel. 0770 2140/2401).

Special Points to Remember: Leaders of geological parties and individuals are warned of the dangers that exist to themselves and their parties when visiting geological sites in the moorland and mountain areas of Arran where game shooting takes place. Geologists are therefore asked before entering any of these areas, at any season of the year, to contact the Tourist Information Centre (as detailed above) who will provide the name, address and telephone number of the Head Stalker. He will advise on local conditions which can change from month to month.

It cannot be too strongly stressed that the principals of conservation should be strictly observed and that hammering should be kept to a minimum. Several interesting localities have been seriously damaged by the inconsiderate actions of badly supervised groups over the years.

Acknowledgements: The simplified physiographical and geological maps of Arran forming Figures 1 and 2 are based on the one-inch Ordnance and Geological Survey maps of the island and due acknowledgment is made in both cases. In the preparation of the geological sketch-maps accompanying the excursions the writers wish to record their indebtedness for permission to consult the standard six-inch maps of the Geological Survey. Figures 8 and 12 are taken from the *Book of Arran,* published in 1910 by the Arran Society of Glasgow and are reproduced by kind permission of the Society.

In the preparation of this revised edition the editors are grateful to Dr. B. J. Bluck for his advice and help in the field, and to Dr. B. G. Curry, Dr. A. E. Mussett, Dr. F. Whyte and many others who have suggested additions and improvements. In the preparation of the figures invaluable help was provided by Mr. D. Maclean, Mrs. S. Hall and Mr. A. S. Wood, and special thanks is due to Professor B. E. Leake for his encouragement and the provision of the facilities of the Department of Geology, University of Glasgow.

2 SUMMARY OF GEOLOGY

With special reference to localities selected for examination

The wide range of geological systems and rock formations represented in Arran is shown in the table on page 21. From this it can be realised why the island has sometimes been said to provide an epitome of Scottish geology. Arran has a rich variety of sedimentary, igneous and metamorphic rocks. Examples within each category can as a rule be readily differentiated in the field with the aid of a hand lens. Sometimes, however, it is necessary to confirm field determinations by examining thin sections under the microscope. In Plate II, facing page 52, a representative selection of drawings of microscope sections, all from Arran localities, has been reproduced in order to illustrate variations in structure, texture and mineral content. The descriptive notes accompanying the drawings were prepared by Dr. J. Phemister. The brief accounts of the various systems given in the paragraphs that follow are intended to link, in a connected narrative form, the subject matter of the different excursions.

1. Dalradian

(*Excursions 7 to 10 mainly*)

The rocks grouped under this name are the oldest in Arran. They form part of a broad belt of metamorphic rocks of diverse types, which stretches across Scotland from Kintyre to the Aberdeenshire coast. The name Dalradian was given to them by Sir A. Geikie in 1891. In Arran they are confined to the northern part of the island where they form a ring round the Northern Granite, interrupted only along a distance of some five km on its eastern margin. The maximum width of the ring is about three km. Rock types present include:

ERA	SYSTEMS AND SUB-DIVISIONS		CONDITIONS OF DEPOSITION	THICKNESS (in metres)	IGNEOUS ROCKS		AGE in m.y.
					CONTEMPORANEOUS	INTRUSIVE	
QUATER-NARY	RECENT AND PLEISTOCENE	Blown sand : Peat Alluvium	Fluviatile				
		Raised beach deposits	Marine				
		Moraines	Glacial				
		Boulder clay	Glacial				1.8
TERTIARY	No sedimentary rocks				Lavas in Central Ring Complex	Granite, Central Ring Complex: Sills and dykes	65
MESOZOIC	CRETACEOUS / JURASSIC	Only as masses and fragments in the Tertiary Central Volcanic Ring Complex					195
	TRIASSIC	Marls and sandstones	Lacustrine: fluviatile	700			230
UPPER PALAEO-ZOIC	PERMIAN	Sandstones & breccias	Aeolian: torrential	200	lava (in conglomerate)		280
	CARBONIFEROUS	Coal Measures	Deltaic	293 (at Corrie)		quartz-dolerite dykes	
		Passage Group (Millstone Grit)	Deltaic		Lavas locally		
		Carboniferous Limestone Series	Marine; estuarine		Lava, agglomerate	a few dykes	
		Calciferous Sandstone Series	Lagoonal				345
	OLD RED SANDSTONE	Upper	Fluviatile	920	Lavas locally		
		Lower	Lacustrine : fluviatile	1200	Lavas locally	a few sills	395
LOWER PALAEO-ZOIC	SILURIAN	Not represented					
	ORDOVICIAN	Arenig	Marine	? 300	Lavas	Gabbro	500
	DALRADIAN (? Cambrian)	North Sannox schistose-grits, schists and slates		1800 +			

TABLE 1. The Geological Systems and rock-formations of Arran.

(*a*) Quartzose schistose grits, varying in texture and composition: originally coarse-grained sandstones.
(*b*) Coarse pebbly grits, quartzose and feldspathic, often in strong massive beds up to about 2 m in thickness; the pebbly bases of some of these carry quartz pebbles up to 10 or even 12·5 cm in length; originally impure pebbly sandstones or conglomerates.
(*c*) Fine-grained schists, including greenish-grey phyllites with abundant chlorite, and dark grey slates: originally shales and sandy shales.
(*d*) At least one band of impure sandy limestone, associated with dark and sometimes graphitic slaty beds; limestone is only recorded, however, at a few localities between Pirnmill and Catacol (Excursion 10, locs. 20–21).

Group (*a*) and (*b*) form the bulk of the Dalradian and since they reach their fullest development in the hilly ground northeast of the Chalmadale–North Sannox valley they have been collectively termed by Anderson (1945, p. 268) the North Sannox Grits.

Succession and structure. In the hilly tract just mentioned, extending from North Newton southeastwards to Corloch, Gunn (1903, pp. 15–16; 1928, pp. 21–22) stated that "there appears to be a regular and apparently ascending series in the schists". He also distinguished in this ascending series ten different sub-groups but it is now known that these rocks in Arran have shared in the complex folding that affected the Dalradian along their whole outcrop and from the observation of graded bedding Anderson was able to establish that the ten groups were in fact six, some of which are repeated across the axis of a fold (Fig. 4). Anderson concluded:

(1) that the North Sannox Grits, subdivided into six distinct groups, lettered A–F on Figure 4, form a continuous upward succession in the ground between Creag Ghlas Laggan and Corloch, with the Loch Ranza Slate Group at the base. Their total thickness may well be of the order of 1,980 metres.

(2) that the Loch Ranza Slate Group occupies the core of a fold

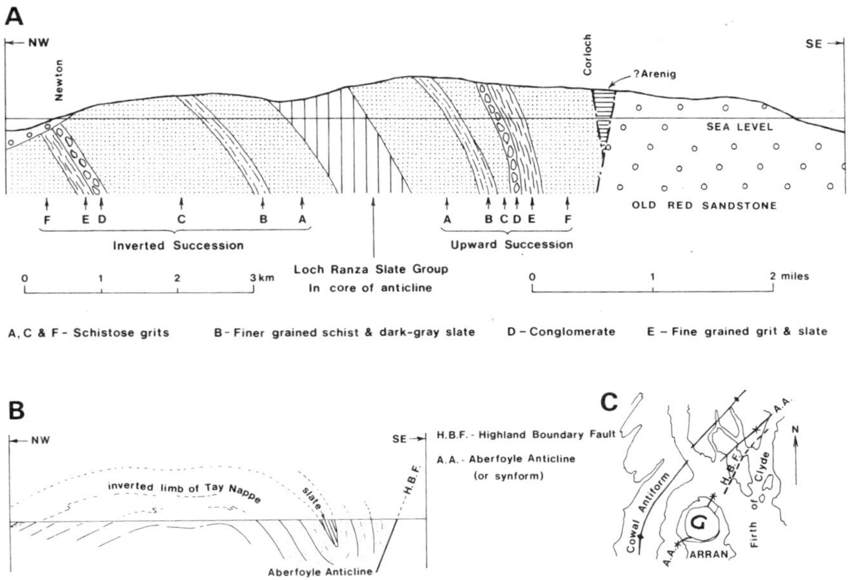

FIG. 4. A, Cross section of Dalradian of North Arran from Newton to Corloch.
B, Structural model of the Tay Nappe according to Shackleton (1958). Note the position of the slate at the core of the Aberfoyle Anticline corresponding with the Loch Ranza Slate Group in section A. Length of section B is about 30 km.
C, Major structures in the Clyde Area. Note how the Aberfoyle Anticline has been distorted by the intrusion of the granite (G).

in which both limbs dip steeply to the southeast—the Loch Ranza Anticline (it is actually a synform or inverted anticline).

(3) that along the overturned northwestern limb of the anticline the sub-groups of the North Sannox Grits are repeated in *inverse* order. Here there are many examples of successive grit bands, varying individually in thickness from 0·6 to 1·8 metres, in which the graded bedding indicates inversion.

The structure shown in Figure 4 is of course simplified. There are, for example, minor folds inclined in the same direction as the major one. Cleavage/bedding relationships clearly show that the fold closure formed by the Loch Ranza slates is downward facing and hence it is a synform (an inverted anticline) closing downwards and not an ordinary anticline as supposed by Anderson. In fact it represents the overturned hinge of a very much larger structure, the Tay Nappe (Shackleton 1958, pp. 361–389). The formations to the east of the slate core belong to the upper limb of the Tay Nappe and young eastwards; those to the west form part of the lower limb. The latter are inverted and young downwards.

In Arran two cleavages (schistosities) are associated with the Tay Nappe and relate to fold episodes. The earlier is associated with highly compressed folds with attenuated limbs and is strongly penetrative. The later is often expressed as a flaggy "fracture" cleavage (or strain slip cleavage) and is associated with folds which are seen most clearly where they form bands of fold hinges along which quartz has been thickened, the limbs being thinned out in the more micaceous lithologies. In the slaty members the two cleavages are coincident but in the quartzitic (psammitic) material the two are at angles of 10 to 20 degrees or so and give the rocks a lenticular appearance.

To the west in Arran a third phase of folding is seen. It can be correlated with the Cowal Anticline, on the mainland to the northeast. The style of folding is concentric and very complex but the earlier fold structures, cleavages and fold hinges, are still clearly discernible.

Catacol Synform. This structural feature, termed a synform rather than a syncline since it affects previously folded schists, was first described by Gunn (1903, p. 15). It results from the doming of the Dalradian schists by intrusion of the Northern Granite in Tertiary times. Along the eastern margin of the granite the southeasterly dip of the schists is steepened, but along the western and northwestern margin the rocks are tilted upwards and thrown into a pronounced synclinal fold, the axis of which crosses the coast at several points in the neighbourhood of Catacol (see Figs. 13 and 14).

Age of the Dalradian. The Dalradian rocks in the vicinity of the Highland Border are, at least in part, of Cambrian age. The age cannot be determined in Arran: the evidence is based on the results of research on rocks of other areas to the northeast, notably the discovery of Middle Cambrian fossils at Callander in limestones and shales interbedded in grits forming the upper part of the Dalradian succession (Pringle 1940, 1942; Anderson 1947, p. 484).

2. Ordovician

The rocks that have been included here are confined to a narrow outcrop which crosses the North Sannox Water a little below the Sannox–Loch Ranza road bridge. The outcrop in the stream is about 300 metres in width and it extends up the hillsides to north and south over a total distance of about 2·8 km (Excursion 6, locs. 6, 14, 15, 16). Gunn (1903, p. 18; Tyrrell 1928, p. 23) described the assemblage of rocks here as "an essential part of the metamorphic series" but from their general lithological resemblance to the Arenig of south Ayrshire classified them as ?Arenig. Anderson and Pringle (1944) recorded the occurrence in them of two poorly-preserved hingeless brachiopods referred to *Acrotreta* and nearly identical with forms in the Arenig of south Ayrshire and Aberfoyle, and concluded that the age of the Arran rocks was definitely Arenig. More recently the discovery of a diverse silicified fauna in a limestone from Aberfoyle has confirmed an Arenig age for at least part of the Highland Border rocks (Curry *et al.* 1982).

Anderson (1947, p. 512) ascribed the metamorphism of Dalradian and Arenig alike to earth movements affecting Cambrian and Ordovician rocks along the Highland Border but not Devonian (Old Red Sandstone), that is, to the Caledonian Orogeny. On the other hand a pre-Arenig age for the metamorphism of the Dalradian is strongly supported by more recent work (Longman *et al.* 1979; Bluck *et al.* 1980; Curry *et al.* 1982). If the Highland Border rocks of Arran are the same age as the limestones and shales at Aberfoyle, then they post-date the Dalradian folding and metamorphism (*ibid.*).

There is no evidence anywhere in the Midland Valley region of a transition from Dalradian (Cambrian) to Arenig, certainly none in Arran where Anderson and Pringle have interpreted the western junction as a folded disconformity. Elsewhere along the Highland

Border the junction between Dalradian and Ordovician is a fault. If the Arenig age of the Arran rocks is accepted on lithological and palaeontological criteria (the latter admittedly slender) both Dalradian and Arenig must have been affected by the later Caledonian orogenic movements.

3. Old Red Sandstone

The Old Red Sandstone, largely owing to the pioneer researches of Hugh Miller (1802–1856), occupies a special place in the annals of Scottish geology. It is noted for its immense thickness, its prevailing red colour and its abundance of massive conglomerates; it is famous also for its fish-beds with their sometimes prolific remains of primitive types of fish and for the evidence it has yielded locally for true land plants, the forerunners of the rich and varied Carboniferous floras. Through the Midland Valley of Scotland the Old Red Sandstone falls into two major divisions, separated by a marked discordance.

3a. Lower Old Red Sandstone

(Excursions 1 and 6 mainly)

With the close of Lower Palaeozoic times a new chapter in the geological record opens. As a result of powerful crustal disturbances known collectively as the Caledonian Orogeny, an entirely new topography had been produced. Scotland became part of an extensive land mass with lofty mountain ranges and broad stretches of intervening lowlands. One of the latter lay along the site of what is now the Midland Valley and it was in this that the Lower Old Red Sandstone sediments and their locally interbedded volcanic rocks slowly accumulated. Where most fully developed their total thickness is estimated to be as much as 5,800 metres. The sediments themselves, sandstones, mudstones and conglomerates, represent widespread flood-plain materials, derived by degredation of the high ground to the north (metamorphic Dalradian) and to the south (closely folded Ordovician and

Silurian) laid down in a slowly subsiding intermontane depression.

The Lower Old Red Sandstone of Arran has an outcrop stretching from near Corloch southwards across North Glen Sannox and Glen Sannox to Glen Rosa and Shurig, thence turning westwards and sweeping across the centre of the island by An Tunna (361 m) and Garbh Thorr (299 m) to Auchencar and the coast south of Dougrie. Along its concave inner margin it is nearly everywhere in contact with Dalradian schists. The boundary line between Corloch and the granite margin near Cnocan Donna may be a continuation of the Highland Boundary Fault which crosses Bute from Rothesay to Scalpsie Bay and if projected to the SW comes to shore in Arran near Corloch; however, if it is the Highland Boundary Fault it does not express itself as clearly as it does on Bute or the mainland to the northeast.

From Cnocan Donna to a little south of the Corrie Burn the Lower Old Red Sandstone is in faulted contact with the Northern Granite along a markedly curved line interpreted by Bailey (1926) as originating with the emplacement of the granite. Along this line at various places there is evidence of crushing and mylonitisation, notably in the Corrie Burn (Excursion 5). It should be noted that to the west of Corrie a narrow strip of Dalradian rocks borders the granite for a distance of about 800 m (Excursion 5, loc. 11a). South of the Corrie Burn contact and along the rest of the outcrop the boundary between the Lower Old Red Sandstone and the Dalradian has been taken as a continuation of the Highland Boundary Fault and is shown as such on geological maps of the island. Gunn himself, however, in his characteristic cautious way expresses doubts on this point (1903, pp. 128–129). In papers dealing in detail with the Old Red Sandstone and the Highland Boundary Fault in Arran (Friend, Harland and Hudson 1964; Friend, Harland and Smith 1970) the Old Red Sandstone–Dalradian boundary has been reinterpreted as a step-faulted unconformity and red-stained Dalradian has been traced around the northern part of the island.

The rock types represented are; conglomerates with generally rounded boulders of quartzite and vein quartz; reddish feldspathic pebbly grits; and purplish to reddish thin-bedded sandstones and siltstones. The total thickness of the formation may reach about 1,200 m. Broad local sequences may be established in different parts of the island, but correlation from one district to another is difficult because of intraformational discordances, local folding and lateral variations in thickness and facies. Good sections of the various sedimentary types can be studied in Glen Shurig and the Allt an Bhrighide (Excursion 1, locs. 14 to 18), in Glen Sannox (Excursion 6, locs. 1 to 5) and along the coast south of Dougrie. Some of the conglomerates contain numerous boulders of andesitic lava, while the associated sandstones carry grains of igneous material showing tiny feldspar laths (Excursion 1, loc. 16 and Excursion 6, loc. 3).

A useful marker horizon on the western side of the island is the interbedded volcanic series, about 25 to 30 m thick, which extends for about 4 km between Auchencar and the northwest side of Garbh Thorr. Both hornblende-andesites and olivine-andesites are present. The former are specially abundant in the conglomerates referred to and their occurrence at different localities would suggest that the volcanic series had originally a wider extension.

Along the southern margin of the Central Ring Complex sporadic exposures of strata referred to the Lower Old Red Sandstone appear in the upper part of Clauchan Glen and at the heads of Benlister Glen and Glen Dubh. At these localities the beds have been indurated and altered by the intrusive masses of the complex. The original red and purple colours of the grits and mudstones have been changed to bluish and greenish-blue tints, while the bedding and other structural features have been largely obliterated.

The semicircular outcrop of the Lower Old Red Sandstone in conformity with the eastern margin of the Northern Granite has already been mentioned. Along the outcrop the dips, generally steep, swing round from NNE, in the northern part to SSE, or even S, in the southern part. This structure, in which the Upper Old Red Sandstone, the Carboniferous and the Permian rocks share, was termed by Gunn (1903, p. 131) the North Sannox

Anticline. He recognised that it was due to uplift produced by the intrusion of the granite.

3b. Upper Old Red Sandstone

During the interval that followed the close of Lower Old Red Sandstone times and preceded the deposition of the Upper Division further crustal movements occurred. The Lower Old Red Sandstone rocks were folded and faulted and then subjected to prolonged denudation. In some areas, indeed, they were completely removed.

In Arran the outcrop of the Upper Old Red Sandstone follows the same circular course round the eastern and northeastern margins of the granite as that of the Lower Division. Along this tract there is no clear structural evidence of any unconformity between the two. In the north of Glen Sannox, however, there is at one point (Excursion 6, locs. 18–19) indications of an angular break along which the Upper Old Red Sandstone appears to overstep different members of the Lower. On the northwestern side of the Boundary Fault near Corloch there is a small outcrop of Upper Old Red Sandstone sediments, about 50 m thick, resting on Dalradian schists and followed upwards by Carboniferous strata (Calciferous Sandstone Series). At North Newton, again, a metre or two of calcareous sandstone, now placed in the upper part of the Upper Old Red Sandstone, directly overlie the upturned edges of the Dalradian (Hutton's Unconformity: see Excursion 7, loc. 4). At both these localities the Lower Old Red Sandstone has been completely overstepped.

In the northeastern exposures a rough three-fold division of Upper Old Red Sandstone is present. The lowest part of the succession consists of about 430 m of thin-bedded reddish or purplish sandstones and massive bands of conglomerate with cobbles and boulders of vein quartz (mainly), quartzite and quartzose schists. Gunn has noted that the quartz fragments give a decidedly pale colour to the conglomerates although the matrix of sand is reddish. Above this a series of red sandstones passes upwards into an alternating conglomerate/sandstone succession, about 490m thick, best exposed along the coast at Sannox Bay. The highest beds in the succession contain irregular, often nodular bands of cornstone and they were formerly taken as forming the basal beds of the Carboniferous. Cornstones occur near the Fallen Rocks and again on the shore a little north of Corrie. It may be that these two separate occurrences are connected by an undersea arcuate outcrop.

Lying between the lower sandstones and conglomerates and the red sandstones mentioned above there is a thin development of basaltic lavas with a maximum thickness of perhaps 30m. The greatest part of the outcrops of these lavas, as mapped by Gunn, are now covered by afforestation and dense undergrowth. Such material as has been collected recently is, as previously described (Tyrrell 1928, p. 43), purplish-red, amygdaloidal and decomposed.

4. Carboniferous

(Excursions 4 and 5 mainly)

The Carboniferous rocks outcrop mainly in the northern part of the island, where they occur in a series of six faulted tracts forming an interrupted semicircle northeast, east and southeast of the Northern Granite. Two of these are coastal and provide good sections; the other four are interior and only partial sections are available.

1. The northeast coast from near the Cock of Arran to near the Fallen Rocks; a strip about five km long and about 0·5km in width; generally referred to as the Laggan section.
2. The east coast at Corrie, a narrow strip about 1·2km long (Excursion 4).
3. The Locherim Burn and adjacent streams at High Corrie (Excursion 5).
4. Sections in the Uisge nan Fear, about 400m south of the Corrie Burn (Calciferous Sandstone Series) and at the old quarry in the Corrie Limestone at An Scriob north of Maol Donn.

5. The Merkland Burn and other localities north of Brodick Castle.
6. Hillside south of the String Road, Glen Shurig (Excursion 1, locs. 19 to 22).

   There is also a narrow fault-bounded strip of Carboniferous rocks (Carboniferous Limestone Series) at the Fairy Dell on the north coast east of North Newton (Excursion 7, locs. 6 to 7).

In the southern part of the island rocks of Carboniferous age occur in a broken and interrupted series of narrow outcrops to the east and southeast of the Central Ring Complex as well as in the tributaries of the Sliddery Water south of the summit of the Ross Road. They are briefly referred to in later paragraphs; exposures are generally poor and disturbed by faulting and intrusions. West of the Central Ring Complex, in the Machrie district, there is no trace of Carboniferous rocks, the Permian resting directly upon the Upper Old Red Sandstone.

Subdivisions and rock-types: The major subdivisions of the Carboniferous represented in the Laggan and Corrie Sections are set out in the table below, together with the approximate thicknesses assigned to each.

			Laggan m	*Corrie* m
UPPER CARB.		Coal Measures Passage Group (Millstone Grit)	91	84
UPPER CARB. / LOWER CARB.	Carboniferous Limestone Series	Upper Limestone Group Limestone Coal Group Lower Limestone Group	229	130
LOWER CARB.	Calciferous Sandstone Series	Upper Sedimentary Group	175	84
		Volcanic Group	114	137
		Lower Sedimentary Group	214	23
		Total thickness of Sediments	709	321

The figures of thickness of the Coal Measures are after Leitch (1942); the others are derived from Gunn (1903, p. 46) but his estimates for the Lower Sedimentary Group include basal beds with nodular cornstones now placed in the Upper Old Red Sandstone. In the detailed account of the Corrie shore section on pp. 103 to 111 the writer has suggested slightly different thicknesses for some of the groups. The Passage Group of the Midland Valley Carboniferous succession was formerly named Millstone Grit but has been shown not to be strictly equivalent to the Millstone Grit of the North of England. The base of the Upper Carboniferous in the Midland Valley of Scotland coincides with the base of the Limestone Coal Group.

The rock types represented in the Carboniferous formations in Arran are described in the notes for Excursion 4: Excursion 1, locs. 19 to 22; Excursion 5, locs. 1a to 6a; and Excursion 7, locs. 6, 7, 11 and 12. The general notes that follow are given to supplement these accounts.

Calciferous Sandstone Series. Both sedimentary groups attain their fullest development on the northeast coast. The rock-types include sandstones and grits, sometimes pale-grey, sometimes mottled with reddish patches, variegated mudstones, and thin bands of a grey muddy cementstone more common in the strata underlying the lavas. The volcanic group consists in the main of olivine-basalt lavas containing phenocrysts of olivine and augite (Craiglockhart type) or microphenocrysts of olivine and occasionally labradorite and augite (Dalmeny type). For a full account of the classification of the olivine-basalts of Carboniferous age in the Midland Valley of Scotland see Macgregor and MacGregor (1948, p. 54). Good exposure of the lavas, with subordinate bands of agglomerate and ash, are available at localities in the Laggan area, at Corrie both on the shore and in the burns draining the hillsides to west and south, and in the Uisge nam Fear. Thin developments of the volcanic rocks and associated sediments are present in the Merkland Burn, on the southeast of the String Road in Glen Shurig (Excursion 1, locs. 19 and 22), and in the centre of the island in the upper part of Benlister Burn, south of Brisderg, and near Cnoc a'Chapuill. In the last mentioned locality any Carboniferous beds exposed belong to the Calciferous Sandstone Series and are overlain unconformably by the Permian.

Carboniferous Limestone Series. This series, noted for its great variations in thickness across major fault lines on the Ayrshire mainland, shows in Arran a marked attenuation from about 230 m in Laggan to a little over 120 m at Corrie. The Corrie Limestone at the base is a recognisable horizon throughout, but any higher limestones present cannot as yet be correlated individually and no definite boundary of the Upper Carboniferous (between the Lower Limestone and Limestone Coal Groups) can be drawn. In the latter at Laggan, coal was at one time worked on the foreshore near Cock Farm, about 1·2 km southeast of the Cock of Arran. According to one old writer, the seam showed coal 25 cm on foul coal 20 to 25 cm, on coal 40 cm. It was used in the manufacture of salt in salt pans situated close at hand, and also for the burning of limestone. The mining here, which must have been on a very small scale, was carried on towards the end of the eighteenth century. There is no coal at Corrie or indeed anywhere else in the island, apart from a few thin ribs of coal in beds underlying the volcanic rocks 800 m or so southeast of Laggan Cottage. Possible equivalents of the Index Limestone, which marks the base of the Upper Limestone Group of the mainland, are present at both Laggan and Corrie, but while at Laggan there are three or four well-developed limestones above the Index, there are only a few thin shelly limestones in the Corrie sequence. In the Brodick Castle area, again, only one higher limestone belonging to the Upper Limestone Group has been recognised, while on the southeast side of Glen Shurig there is no trace of the Upper Limestone Group and the whole Carboniferous succession may not exceed about 150 m. Further south, in the small detached outcrops at the head of the Benlister Burn (south of Brisderg and east of Cnoc na Croise) the Lower Carboniferous sequence consists of white sandstone, a metre or so, on calcareous shales and limestone (the Corrie

Limestone), on red mudstones overlying a thin development of volcanic rocks, all successively transgressed by Permian grits. This attenuation of the Calciferous Sandstone Series and Carboniferous Limestone Series may be attributed to differential crustal movements giving rise to conditions in which sedimentation was greatest in the northeast and thinned southwards towards a land area over south Arran (see George 1960, pp. 74 and 76).

Passage Group. The beds described by Gunn as Coal Measures are now known to include, in some localities, rocks referable to the Passage Group (formerly referred to as "Millstone Grit"). In the Merkland Burn, about 1·2 km north of Brodick Castle, red mudstone and marls, 22 m thick and with an ironstone containing mussels at the top, closely overlie beds which include at the base a clay of bauxitic type, resting upon decomposed red ash and lava. These volcanic rocks and the overlying bauxitic clays are correlated with the Passage Group Lavas and Bauxitic Clay of north Ayrshire. Leitch (1942, p. 145) has recorded the occurrence of bauxite-like clays associated with volcanic ash in the Benlister Burn 4 km west of Lamlash. He has also suggested that the volcanic rocks occurring below red shales containing mussels in one of the head streams of the Sliddery Water on the south side of the Ross Road may be of Passage Group age. It may be noted that in the Machrihanish coalfield the Passage Group consists of a succession of lava flows, with some sedimentary intercalations, about 148 m thick. No volcanic rocks are present at this horizon at Corrie or at Laggan, although the occurrence of sedimentary rocks of Passage Group age cannot be excluded.

Coal Measures. The type of sedimentation characteristic of the Coal Measures of Arran can be studied on the Corrie shore (Excursion 4, loc. 11) and at the northwest end of the Laggan section (Excursion 7, locs. 11 and 12). Details of the section at Corrie, as worked out by Leitch, are given on page 110; the same author has also given a detailed description of the same beds at Laggan (1942, p. 149). Disturbed and contorted bedding is a common feature in the sandy beds, especially at Laggan. Variegated rooty seat-earths occur, sometimes overlain by reddish shales, but there are no traces of coal. Nevertheless the mussels obtained at various localities have been sufficient to indicate their position in the non-marine lamellibranch succession established for the mainland. Leitch's tentative conclusions are set out diagrammatically below.

Faunal Zones	*Laggan*	*Corrie*	*Merkland Burn*	*Sliddery Water*
Lower *similis-pulchra* zone	Absent	Absent	×	×
modiolaris zone	? ×	×	Absent	Absent
communis zone	?	?	Absent	Absent

Each of these faunal zones in the mainland succession includes valuable coals. The diagram above suggests, as Leitch pointed out, that between the Passage Group volcanic rocks and the lower *similis-pulchra* zone of the two southern localities there is an unconformity cutting out the *communis* and *modiolaris* zones, while in the Merkland Burn and at Sliddery Water head, the sediments yielding *similis-pulchra* forms underlie gritty sandstones and lavas both placed by Leitch in the Permian (p. 39).

5. New Red Sandstone

The sedimentary rocks that overlie the Carboniferous were originally grouped under the general designation of the New Red Sandstone in contradistinction to the underlying Old Red Sandstone. It was recognised later, however, that they included two distinct rock divisions, differing in their lithology and mode of origin. Gunn (1903, p. 67) suggested a division into Upper and Lower Trias whereas Tyrrell (1928, p. 76), purely on the basis of lithology, preferred to consider the lower part of the succession, consisting of sandstones and breccias, to be Permian, and the upper, generally finer-grained division with its higher proportion of siltstones and marl, to belong to the Triassic Period.

The Carboniferous Period had been one of recurring shallow seas, widespread estuaries, broad deltaic flats and swamps. By contrast the deposits of the succeeding Permian are wholly continental and initially aeolian (wind-blown) or torrential in origin. Between the two periods far-reaching geographical changes were brought about by powerful earth movements which began in late Carboniferous times and continued into Permian times. These crustal disturbances, known as the Hercynian orogeny, elevated and folded the Carboniferous rocks and brought them within the reach of the agencies of erosion. It was upon this uplifted Hercynian tableland—a land-locked area with internal drainage—that the desert sandstones and breccias of the Permian were laid down. Although now found only as residual outcrops they no doubt at one time covered much of the Midland Valley region.

Doubt as to the Permian age of the lower part of the succession was raised by the discovery of plants of apparent Upper Carboniferous (Westphalian D) age in a band of sediment intercalated between lava flows of presumed Permian age, in an exposure in the River Ayr near Stair (Mykura 1960). Subsequent radiometric age dates (de Sousa 1982, p. 460), however, establish the Permian age of the lavas and hence that of the overlying dune bedded sandstones of the Mauchline Basin of Ayrshire, and by inference, the lithologically similar beds of Arran.

The stratigraphical divisions within the New Red Sandstone of Arran were originally made on a purely lithological basis by comparison with the Permian and Triassic rocks of the Midlands of England. Thus it was considered (Tyrrell 1928, p. 76) that in the succession that can be followed southwards from Corrie to the south of the island all the beds that are dominantly composed of sandstones should be identified with the Permian and only the uppermost part of the succession, within which marls and shales become important, should be assigned to the Triassic. Following this logic all the beds that outcrop round Lamlash Bay were included in the Permian and the boundary with the Trias was put in a position just south of Largybeg (Excursion 12, loc. 2).

Craig (1965, pp. 387–388) emphasised the variability of the

nature of the deposits within the succession and the considerable interlensing of strata that makes it difficult to correlate beds over quite short distances and suggested a simple three-fold division of the New Red Sandstone of Arran without attempting to place the position of the Permian/Trias boundary. The subsequent discovery of identifiable microfossils has allowed a refinement of Craig's succession (Warrington 1973, p. 110) suggesting that a large part of the Lamlash Bay succession should be included with Triassic rather than the Permian. The major stratigraphic subdivisions can be summarised as follows:

	Stratigraphic divisions according to Warrington (1973, after Craig 1965)	*Subdivisions of Tyrrell (1928, pp. 76–77; after Gunn 1903, p. 67)*
		7. Levencorroch Marls and Cornstones
	Auchenhew Beds (c. 300 m)	6. Auchenhew Sandstones and Shales
TRIASSIC		5. Lag a'Bheith Marls and Cornstones
	Lamlash Beds	4. Glen Dubh Sandstone
———?	(c. 400 m)	3. Lamlash and Machrie Sandstone
PERMIAN	Brodick Beds	2. Brodick Breccia
	(c. 200 m)	1. Corrie Sandstone

The precise position of the division between the Permian and Trias remains uncertain but is most likely to be within, or near the base of the Lamlash Beds.

5a. **Permian** (Brodick Beds)

(Excursions 1, 2 and 13 mainly)

The rocks included in this system appear at the surface along a coastal strip of varying width between Corrie and Brodick. South of Brodick they continue along the coast towards Clauchlands

Point but by the time Lamlash Bay is reached the rocks form part of the Triassic System. Inland they occur in Glen Cloy. On the west side of the island Permian rocks outcrop along the coast north and south of Machrie Waterfoot and in the north an isolated outlier of sandstones and breccias occurs at the Cock of Arran. At this latter locality pebbles of Carboniferous limestone yielding marine shells were first obtained in the basal beds over one hundred years ago (see page 130). The most complete sections occur on the east side of the island where contrasting lithologies can be studied at and south of Corrie (Corrie Sandstone) and in the vicinity of Brodick (Brodick Breccias).

(1) **Corrie Sandstone.** This group of beds, consisting essentially of a thick series of brick-red, markedly false-bedded sandstones, can be best examined along the shore between Corrie and Brodick and in the old quarries above the raised beach. The desert origin of the beds is shown by the characteristic dune-bedding, well seen, for example, on the north shoulder of Maol Donn, and by the rounded "millet seed" quartz grains (Plate II, Fig. 1).

(2) **Brodick Breccias.** These beds are made up of a succession of breccias or conglomerates associated with massive, red, dune-bedded sandstones of Corrie Sandstone type. The fragments in the breccias, sometimes angular, sometimes rounded, are set in a matrix of wind-rounded and wind-polished sand grains. The fragments themselves, which vary greatly in size, are dominantly of quartzitic rocks—vein quartz, quartzites, quartzose schistose-grits—but other types occur such as basalts of Carboniferous affinites and, near the top of the succession, agate pebbles. Good exposures of the Brodick Breccias occur along the coast south of Brodick, in the central part of the Benlister Burn, in Glen Dubh and at the Cock of Arran.

Correlation and Condition of Deposition. The Corrie Sandstone is markedly similar in facies to the Mauchline Sandstone of central Ayrshire, a correlation suggested by Gunn (1897, p. 273) and by Geikie (1897, vol. 2, pp. 58, 418, 419). In the Mauchline Basin the red sandstones have a maximum thickness of 311 m and are underlain by a volcanic group consisting of lavas and lava-breccias which may reach a thickness of over 240m. In Arran, however, volcanic rocks are almost entirely represented by basalt fragments present in the upper part of the Brodick Breccias. A 9m band of slaggy amygdaloidal lava in one of the head burns of Sliddery Water on the south side of the Ross Road, about 5km southwest of Lamlash, was tentatively assigned to the Permian by Leitch (1942, p. 147) and is also so lettered on the latest edition of the 1:50,000 geological map. It is overlain by red sandstones showing the characteristic wind-rounded grains of quartz, some of which have infiltrated into cracks in the upper surface of the lava. This lava was included by Gunn and Tyrrell in the Coal Measures.

The evidence that the Permian in Arran rests upon different zones in the Coal Measures has already been indicated (p. 35). In the upper part of the Benlister Burn it overlies the Corrie Limestone or the shales associated with the limestone, whereas farther west, in the Machrie district, it rests upon Old Red Sandstone sediments. It is thus markedly transgressive across a variety of older rocks. Where its junction with the Old Red Sandstone is seen in Machrie Bay there is a transitional zone of broken and badly weathered material suggesting a land surface.

The conditions under which the Corrie Sandstone and the Brodick Breccias appear to have accumulated have been mentioned already. However, rather than thinking of these beds as two distinct stratigraphic units it would probably be more accurate to treat them as two different types of sedimentary rock which could have been deposited at the same time in different parts of the Permian desert floor. There is evidence that the breccias and the dune-bedded sandstones intertongue and hence that the whole of the Permian succession should be treated as a single stratigraphic formation named the Brodick Beds (Tyrrell 1915; Craig 1965, p. 387). This was deposited on the edge of a desert basin flanked on the northwest by the ancient Dalradian mountains, and their foothills of Old Red Sandstone and Carboniferous rocks uplifted by Hercynian folding. Much of this desert basin was covered by stony deposits (the Brodick Breccias) washed in by occasional flash floods from wadis in the high ground. In the intervals between deposition the sediments were reworked by the wind, the finest grained material being blown out of the region altogether while the sand accumulated in isolated groups of dunes, a set of conditions that can be seen at the present day in many desert areas such as the Sahara.

5b. Trias

(Excursions 12 and 13)

The Permian in Arran is succeeded by a thick series of sedimentary rocks including all or part of the Lamlash Beds and the entire sequence of the Auchenhew Beds. The identification of this succession with the Triassic System rests on the recognition of microfossils of late Early or early Middle Triassic age in the lower part of the Auchenhew Beds (Warrington 1973) and the occurrence of black shales of Rhaetic age on top.

(1) **Lamlash Beds.** The lower part of this succession is composed mainly of thin-bedded, generally soft, medium-grained sandstones of a reddish or purplish colour. Interbedded with these occur occasional pebbly or conglomerate bands, while locally bands of red mudstone or sandy mudstone showing sun-cracked and ripple-marked surfaces are present. Good sections are available along the Clauchlands shore, at various localities between Lamlash and Whiting Bay, in the Lag a'Bheith, in Monamore Glen and in the Machrie District. Some of the more massive sandstones have been locally quarried for building stone, for example at Lamlash.

The upper part of the succession, referred to by Tyrrell (1928, p. 77) as the Glen Dubh Sandstone, is formed by a series of massive, variegated and often calcareous sandstones. They are exposed on the southern slopes of Glen Dubh above its junction with Glen Ormidale, but probably the best sections are seen at Largybeg where the weathering out of the calcareous cement has given rise to an irregularly channelled and pitted appearance on the rock surfaces. Similar sandstones also occur in exposures on the west side of the island—in Ballymichael Glen and in the Machrie Water roughly 1·6 km from its mouth.

(2) **Auchenhew Beds.** The Lamlash Beds are succeeded by a thick series of sedimentary rocks which in their lithological characters present a strong contrast to those lower in the New Red Sandstone of Arran. They include variegated, often green-spotted shales and marls, fine-grained sandstones mottled in grey and reddish tints and sometimes calcareous, together with frequent bands of pale-coloured, nodular impure limestone or cornstone. These beds were originally referred to the Trias on account of their general similarity to the Trias of the North of England. Their total thickness in Arran is of the order of 300 m or more. It is considered by Warrington (1973, p. 114) that the Auchenhew Beds represent an almost complete sequence of the Middle and Upper Trias: this, by implication, would indicate a Lower Trias or Bunter Sandstone age for much or all of the Lamlash Beds.

The Auchenhew Beds form the country rocks over a large area in the southern part of Arran and many excellent shore and cliff exposures are available to the student along the coast from near Largybeg Point to a little north of Drumadoon Point. Good exposures can also be seen in many of the streams draining the central plateau, for example, in the northern headwaters of the Glenashdale Burn and in the Allt Mòr, the Levencorroch Burn and the Sliddery Water, all reaching the sea along the south coast.

Evidence for the depositional environment of these beds is present in the form of numerous ripple-marked surfaces seen on some of the sandy beds, the shrinkage cracks in the bands of mudstone, the presence in some of the marls of tiny rhomb-shaped crystals of calcite, and the occurrence of layers of nodular calcareous segregations or cornstones all suggesting deposition in sheets of shallow water subject to alternate flooding and desiccation. Deposition in a shallow water fluvial-lacustrine environment appears to be supported by the work of Pollard and Lovell (1976) who studied burrows and associated sediments on the west coast of Arran south of King's Cave but other work (Pollard and Steel 1978) recording halite pseudomorphs, symmetrical ripple marks and other sedimentary structures in the same area, suggests that at least part of the succession was deposited under intertidal conditions adjacent to an arid coast.

With the close of the Triassic Period we come to the end of the succession of solid sedimentary rocks that have a widespread outcrop in Arran. If the earlier chapters in its geological history are, as we have seen, incomplete, there are on the other hand only a few pages of the later chapters extant. Such information as can be gleaned from them must be supplemented by inferences for which there may be few guiding facts. However, by a remarkable and fortunate chance we do possess evidence to show that at one time Arran was covered not only by the red measures of the Trias but also by fossiliferous marine sediments of Jurassic and Cretaceous times. This evidence is outlined in the following paragraphs.

6. Mesozoic Fragments in the Central Ring Complex

One of the most striking results of Gunn's survey of Arran was his discovery of masses of Mesozoic sediments incorporated in the agglomerate of a Tertiary volcanic vent (Gunn 1903, pp. 73–78). These masses include sedimentary types which on lithological or palaeontological evidence can be assigned to the Triassic, Jurassic and Cretaceous Systems. Opportunities to examine some of these in detail are provided on Excursion 14. Green and greenish-grey marls, closely resembling the "Keuper Marls" seen *in situ* at many South Arran localities, form a large mass of steeply dipping, fault-fractured sediments exposed on Muileann Gaoithe at the head of Glen Ormidale 3·2 km west-southwest of Brodick. The Jurassic (Rhaetic and Lias) and Cretaceous fragments, dated on palaeontological evidence, are mainly found on the northwestern edge of the complex within a roughly triangular area bounded by lines joining Glenloig on the String Road and Dereneneach on the Shisken Road with Ard Bheinn. Black shales and thin limestones of Rhaetic age are exposed in the Allt nan Dris, 400 m northeast of Dereneneach Farm. The Lias is represented by calcareous shales and limestone seen in a gully 800 m or so south of Ard Bheinn and the Cretaceous by blocks of pale-coloured limestone carrying nodules of chert exposed, for example, at the "Pigeon Cave" on Ard Bheinn and again on Creag na Fheidh 450 m or so south of Glenloig. These and other masses and fragments of Mesozoic strata represent portions of the original cover of sedimentary rocks through which the Tertiary volcano erupted, and have been preserved by incorporation in the crater materials. Their occurrence shows: (1) that the Trias had a wider extension than its present limits; and (2) that Jurassic and Cretaceous were also represented in the cover rocks of the island immediately prior to the volcanic episode. In this connection it must be remembered that Triassic, Jurassic (Rhaetic to Kimmeridgian) and Upper Cretaceous sediments occur in detached outcrops in the Inner Hebrides (Skye, Raasay, Mull, etc.) and on the adjoining western seaboard (Morvern, Ardnamurchan), where they have been locally protected and preserved under a cover of Tertiary basalt lava. Mesozoic rocks, accordingly, must have extended over a wide area in the west of Scotland, including Arran, and may well have been at one time present over much of the Scottish mainland, for along the east Sutherland coast, between Golspie and Helmsdale, Mesozoic sediments are present in a broken succession ranging from Trias to Kimmeridgian. The Hebridean evidence shows that the Upper Cretaceous beds are transgressive across a denuded surface of Jurassic, Triassic and older rocks. While there is no observational evidence now extant, it is a reasonable inference that the Upper Cretaceous sea covered the greater part of Scotland and that its chalk beds had a transgressive relationship to remnants of Trias and Jurassic sediments similar to that seen in the Hebridean region and in Northern Ireland.

7. Tertiary Period

The regional uplift that followed upon the close of Cretaceous times brought about great geographical changes, with the slow emergence of a new continental landmass of varied relief. These changes, which must have occupied a prolonged interval of time, mark the transition from the essentially marine conditions under which the sediments of the later Mesozoic periods were laid down to the continental conditions which ushered in the succeeding Tertiary Period.

The duration of the Tertiary Period is estimated at approximately seventy million years. It comprises five geological epochs named from below upwards—Palaeocene, Eocene, Oligocene, Miocene and Pliocene. So far as Great Britain is concerned sedimentary rocks of Tertiary age (Palaeocene to Pliocene) are almost entirely confined to southern and southeastern England. In Scotland, on the other hand, sedimentary rocks of Tertiary age are limited to a few sparse occurrences of terrestrial deposits associated with the lavas in the western Highland and Islands.

The broad sequence of events in the story of the Tertiary Period in Arran may be briefly reviewed under the headings: (*a*) interval of early-Tertiary denudation; (*b*) interval of Tertiary igneous activity; and (*c*) interval of late-Tertiary denudation during which the main features of the present landscape were slowly evolved.

7a. Early Tertiary Denudation

The emergence of a new continental land at the end of Cretaceous times, to which reference has been made, was followed by a period of degradation. Much of the cover of Cretaceous and older Mesozoic strata was locally removed and portions of the harder underlying Palaeozoic rocks re-exposed. So much can be inferred from the fact that in the Hebrides and adjacent areas the earliest lavas of the succeeding volcanic episode are transgressive over a wide range of formations from Cretaceous downwards. In western Scotland, as Richey has pointed out (1961, p. 6), where almost every formation of pre-Tertiary age is represented, all of "the rocks concerned belong to the 'floor' over which the Tertiary basalt lavas were extruded".

7b. Tertiary Igneous Episode

Several criteria have been used as guides in attempting to define more precisely the age of the Tertiary igneous rocks. Plant remains obtained from sediments locally associated with the lavas in Mull, Skye, etc., have been regarded as comparable with the flora yielded by Lower Tertiary strata in Greenland and other Arctic regions. The West Highland succession has long been considered, accordingly, to be of early Tertiary age.

Only recently has reliable laboratory evidence begun to be available to set more precise limits to the age of Tertiary igneous activity in Arran. An *Ar* incremental heating date of 60·4±0·6 million years (Evans *et al* 1973) agrees precisely with the Rb-Sr whole rock age of 60·3±1·6 million years (Dickin *et al* 1981) for the Northern Granite, while dates obtained by the former method for the Drumadoon sill, the Holy Island sill and a tholeiite dyke at Port na Feannaiche indicate a narrow range of dates from 58·3 to 56·3 million years (A. E. Mussett 1982, personal communication). A study of palaeomagnetic polarities (Dagley *et al* 1978; Hodgson 1981) supports the evidence that the igneous activity in Arran was confined to a rather brief episode during early Tertiary (late Palaeocene to early Eocene) times and comparisons with both palaeomagnetic and radiometric data from Ardnamurchan, Rhum, Eigg, Canna, Mull and Skye suggests that throughout the Tertiary igneous province in Scotland the activity took place between about 58 and 59 million years ago (Dickin 1981; Mussett 1980; etc.).

Approximately 42% of the area of Arran is occupied by igneous rocks of Tertiary age. These represent a small portion of a volcanic province which at one time must have extended from northern Ireland and western Scotland over much of the continental margin to include St. Kilda and several submerged centres on the continental shelf. "This", in Geikie's words (1910 p. 20), "was the last, and perhaps the grandest, of all the volcanic scenes in the history of the British Isles." Of this old volcanic land only remnants now remain—in Antrim, the Inner Hebrides, Ardnamurchan and Arran.

The Tertiary igneous rocks of Arran fall into the following categories: (1) the numerous sills in the southern half of the island; (2) the Northern Granite; (3) the Central Ring Complex; and (4) the swarm of more or less vertical dykes. Elsewhere in the Tertiary volcanic districts of Scotland one of the earliest manifestations of igneous activity was the outpouring of flow after flow of lava to thicknesses of as much as 2,000 m or more (Mull, Skye, etc.). In Arran there is no direct evidence for a corresponding extrusive phase although it is probably represented by the 2·4 km long ourcrop (or screen) of basic lavas occurring in the Central Ring Complex to the west of Ard Bheinn (see p. 50 and Excursion 14, loc. 5).

(1) **Sills of south Arran.** There were several phases of sill intrusion during the Tertiary igneous activity in the island each with its own characteristic assemblage of rock types. The following broad categories have been recognised:

(*a*) A group represented by sheets of analcime-olivine-dolerite or gabbro, usually referred to in the literature as crinanite although the term is not strictly applicable. The main localities are the Clauchland Hills, Monamore Glen, Kingscross, Dippin, Columbkille and Auchareoch. The three first mentioned are arranged in a broken semicircle around Lamlash. Tomkeieff (1961, pp. 6–8 and fig. 3) put forward suggestions regarding the relationship of these and other intrusions to a hypothetical centre under Lamlash Bay. As a group they are notable for their transgressive relations to the sedimentary rocks in which they are emplaced; for the development of tachylitic, basaltic and doleritic types, often with complicated mutual relationships in their margins and against included masses of sedimentary rock; and for localised severe, pyrometamorphism of the country rocks. From their similarity in petrographic character to olivine basalt of Plateau type they have been regarded as the earliest of the intrusive rocks and to be of the same age as the basaltic lavas preserved in the Central Ring Complex. Although the members of this group have so far been regarded as of fairly uniform composition, it has been found that a thin layer of picrite forms part of the Monamore sheet. It is estimated that the thickness of the larger bodies may have reached 100 m; in the case of the Dippin intrusion a thickness of 42·8 m has been proved by drilling (Gibb and Henderson 1978b). A thorough description of the petrography of the latter intrusion (*ibid.*) accords the most complete investigation of a member of this group to date.

(*b*) A group represented by the sills of quartz-dolerite and associated felsic rock types so abundant in the south of Arran. They vary in thickness from under a metre to 80 m. More or less concordant with the bedding of the country rocks they give rise to well-marked scarp and terrace topography (unfortunately now largely obscured by afforestation). A few transgressive dyke-like bodies occur. Many of the intrusions forming this group are composite, having relatively mafic interior components flanked by more felsic bodies. Occurrences of basaltic xenoliths in doleritic, granophyric or felsitic matrices are numerous. Unlike the analcime-bearing intrusions, they seldom carry xenoliths of country rock. Sills belonging to this group occur in Glen Ashdale (Excursion 11a) and cap many of the tabular hills to the west such as Auchenhew, the Levencorroch Hills and Cnoc Ballygown. Proof that this group is younger than group (*a*) is given in Excursion 12 (loc. 3). The intrusions forming the Tighvein complex are co-magmatic with those of group (*b*), differing mainly in being more coarsely crystalline (Tyrrell 1928, pp. 133–135; Herriot 1975, pp. 7–11).

(*c*) A group of acid sills represented by the massive quartz-porphyry intrusions of Bennan Head (Excursion 12), Brown Head and Drumadoon (Excursion 13). Dun Dubh in the Corrygills district (Excursion 3, loc. 5) is another large intrusion of the same type, probably plug-like in form. A quartz-porphyry dyke 3·35 m thick, occurring on the Corrygills shore east of Dun Fionn may perhaps be connected with it (Excursion 2, loc. 14a). Sometimes the quartz-porphyries are associated with basic rocks, generally tholeiitic, in composite sills. Both the Bennan and Drumadoon intrusions have basic selvedges of this type. A notable feature of these marginal tholeiites is that they usually contain xenocrysts of quartz and orthoclase.

In addition to the generally massive sills listed above there is a group of minor acid intrusions which include the pitchstones and associated felsites for which Arran has long been famous. They may occur as separate intrusions, as composite felsite-pitchstone sills or dykes, or as composite dykes of pitchstone and tholeiite. While many of the intrusions included in this group are sill-like in form and concordant with the bedding of the country rocks, others are dyke-like and transgressive. Sometimes, again, individual intrusions change in form from a sill to a dyke. The pitchstones vary in colour from greenish to almost black and have a resinous lustre. On closer examination they show a glassy base more or less crowded with crystallites which build up the fern-like growths present in many types (Plate II, fig. 7). Tilley (1957) has identified the dominant mineral in these growths as biotite. Some types of pitchstone carry porphyritic crystals of quartz, feldspar (both with inclusions of glass), augite or olivine, the nature of the phenocrysts varying with the rock type.

A number of intrusions of pitchstone or of composite pitchstone-felsite have been chosen for examination during the excursions. Notable examples are the sills on the Corrygills shore (Excursion 2, locs. 11a and 15a); at South Corrygills (Excursion 3, locs. 2 and 3); in the Lag a'Bheith burn (Excursion 3, loc. 12) and at West Shurig Farm (Excursion 1, loc. 23). The well-known composite dykes on the Tormore shore north of Blackwaterfoot may also be included in this list (Excursion 13).

The age, or ages of the various intrusions which comprise this group have not been determined with any certainty. Although they would seem to be later than the members of groups (*a*) and (*b*), they nevertheless probably cover a wide time-range. Some can be referred to the period of intrusion of the quartz-dolerite sills (see Tyrrell 1928, pp. 137, 210), but others may belong to an early stage in the final phase of dyke injection (p. 51). Intrusions of pitchstone, of felsite or of composite pitchstone-felsite are known to cut the great crinanite sheets (Excursion 3, loc. 6), the Northern

Granite in many places (eg. Excursion 6, loc. 8) and the granite of the Central Ring Complex. The felsite of Torr Righ Mòr and Torr Righ Beag, north of Blackwaterfoot, appears to be later than the basic dykes of the dyke swarm, since the dykes cutting the country rocks round it do not transect the felsite itself. The same relationship holds in the case of the riebeckite-trachyte mass forming Holy Island (Excursion 11b).

(2) **The Northern Granite.** This granite mass, dome-shaped and nearly circular, covers an area of some 90 sq. km. It consists of an outer coarse-grained granite and an inner fine-grained variety, the latter intrusive into the former and giving rise to lower and less rugged ground. The minerals comprising both varieties are quartz and feldspar (mainly orthoclase) together with a varying but often meagre content of black mica (biotite). A plutonic intrusion of deap-seated origin, the granite's thick cover of Dalradian and younger rocks has long since been removed by the agencies of denudation. There are three types of contacts or junctions to be examined:

(*a*) The interior junctions of the two varieties of the granite itself. These have been described by Flett (1942) and opportunities for examining some of them are provided during Excursion 9.

(*b*) The intrusive contacts of the granite with the surrounding schists. These can be examined at a number of localities: Glen Rosa (Excursion 1, loc. 13); in the Allt a'Chapuill (Excursion 5, loc. 11a); in North Glen Sannox (Excursion 6, locs. 10, 12 and 13); at Madadh Lounie on the east side of Catacol Glen (Excursion 8, loc. 15a) and elsewhere.

(*c*) The faulted contact of the granite with the Lower Old Red Sandstone in the Corrie Burn (Excursion 5).

Evidence for the forceful nature of the intrusion of the Northern Granite can be seen on its western side where there has been uplift of the lower inverted limb of the Tay Nappe, in the north where the uplift is responsible for swings in the trends of the Dalradians, and along the southern margin where there has been considerable uplift and steepening of dips even in the Permian.

On a number of the excursions there are opportunities for examining the joint systems and mode of weathering of the granite massif. Druses or drusy cavities lined with crystals are not uncommon and occur in both the coarse and fine-grained types. These crystals are in the main the same as those of the enclosing rock, viz. quartz and feldspar. Occasionally, however, translucent crystals of quartz are present, tinted reddish-brown (cairngorm or smoky quartz) purplish (amethyst) or yellowish (Scotch topaz); blue beryl and topaz have also been recorded (Heddle 1901; McCallien 1937).

(3) **The Central Ring Complex.** (Excursions 14a and 14b)—This designation is given to the large oval-shaped volcanic centre lying south of the String Road where Gunn discovered broken masses of Mesozoic strata and so determined the Tertiary age of the igneous episode. This volcanic centre, measuring about 5·6km in an east-west direction and about 4km north to south, cuts Old Red Sandstone strata (chiefly Lower Old Red) along the greater part of its periphery. In and near Glen Ormidale in the east, and between Dereneneach and Ballymichael Glen in the west the complex is faulted against New Red Sandstone sediments. Along its northern margin it clearly cuts the Old Red Sandstone which dips fairly steeply southeastwards as the result of the previous intrusion of the Northern granite. The Old Red Sandstone again dips steeply on the southern and southeastern margins of the complex perhaps indicating a structural parallel with the Northern Granite. To the east and west the marginal relationships are intricately intrusive, faulted or hidden.

The rocks forming the complex are igneous and pyroclastic. Amongst the former group are found intrusions of granite, granophyre, diorite and gabbro (relatively coarse-grained) and finer-grained felsite, porphyry, trachyte and dacite. Extrusive types also occur including andesites and basalts. The pyroclastic rocks are highly varied in texture and composition as they contain fragments of both sedimentary and igneous origin and include both agglomerates and tuffs. Brecciated andesitic and basaltic lavas also occur. Included in the agglomerates are blocks of Mesozoic sediments some of which are of impressive dimensions (page 42). Much finely broken and comminuted debris, including material derived from the Old Red Sandstone, is also included, all probably representing portions of the original cover of country rocks which have been disrupted and shattered by explosive action during the formation of a caldera.

Inspection of the published maps shows arcuate patterns in the outcrops of the igneous and pyroclastic rocks. This general similarity of the Arran centre to the more perfect ring complexes of Mull and Ardnamurchan led Tyrrell (1928, p. 165) to replace Gunn's earlier name of "volcanic vent" by "central ring complex".

For further details of this interesting area the reader is referred to Tyrrell (1928), King (1954, 1955) and Ashford (1958). The median area, from Cnoc Dubh by a'Chruach and upper Glen Craigag to Beinn Bhreac has not received detailed examination.

(4) **The Arran Dykes.** The dykes, by reason of their abundance and their variety in character, are among the most remarkable features in the geology of the island. Collectively they are known as the Arran Dyke Swarm. Here, as in the case of the sills, age relationships are often very uncertain. Some of the larger dykes of analcime-olivine-dolerite no doubt belong to the period of intrusion of the great sheets of the same composition (p. 46). Other dykes, again, may be assigned on petrographic evidence to the time of intrusion of the quartz-dolerite sills (p. 47), or to that of the quartz-porphyry group (p. 47). But the great majority of Arran dykes, mainly basaltic in character, cannot be assigned to any particular stage in the development of the igneous activity. It appears likely that they span three magnetic polarity periods lasting over three million years (Dagley, Mussett, Wilson and Hall 1978; Hodgsen 1981) with the majority being reversely magnetised.

In thickness the Arran dykes vary from about 0·3m to 30m though in the case of one or two composite intrusions the thickness may reach about 60m. The general direction is NNW or NW. In

EXPLANATION OF PLATE II

Fig. 1. Permian sandstone, old quarries at Corrie. ×12.
Grains of quartz, felspar and quartzite, the surfaces of which have been rounded by wind abrasion ("millet seed" grains), are coated and cemented loosely by limonite.

Fig. 2. Lower Carboniferous Sandstone, south side of String Road, 915 m SW of Brodick Church. ×13.
Angular grains of quartz (dusty with fluid inclusions) and of decomposed felspar and siliceous rock are loosely cemented by clay.

Fig. 3. Lower Old Red Sandstone, shore cliff, 800 m S of Dougrie. ×11·5.
Angular and unsorted grains of quartz (clear), decomposed felspar (grey) fragments of mudstone, igneous rocks and oxidized iron ore are compactly cemented by the fine-grained waste of similar material.

Fig. 4. Pebbly grit, Dalradian, near SE end of Creag Ghlas Laggen, North Arran. ×10·5.

Fig. 5. Cleaved grit, Dalradian, shore 69 m ESE of Loch Ranza pier. ×9·5.
The rock has been sheared. Quartz and quartzite pebbles have been deformed and ground away until their long axes lie parallel to the schistosity which is strongly developed by parallel orientation of the chlorite and muscovite flakes of the matrix.

Fig. 6. Slate, Dalradian, old quarries on hillside 2·4 km E of head of Loch Ranza. ×13.
The paler and darker bands represent more silty and more clayey alternations of the original strata. Within the paler bands the effect of shearing can be seen in the development of a lenticular schistosity on a microscopic scale.

Fig. 7. Pitchstone dyke, Schoolhouse, Brodick. ×12.
Small euhedral crystals of quartz lie in a matrix of rock-glass from which numerous crystallites of pyroxene have grown. In other parts the rock contains euhedral prisms of zoned plagioclase and pyroxene which are not shown in this figure.

Fig. 8. Granite, Glen Rosa. ×12.
The rock is composed of oligoclase (showing straight cleavage lines), orthoclase (turbid), quartz (clear, with tiny inclusions and cracks), and biotite (dark, with straight close cleavage). A small prism of zircon lies on the left side of the topmost biotite).

Fig. 9. Olivine basalt lava, Lower Carboniferous, Corrie shore opposite school-house. ×10·5.
Euhedral phenocrysts of purplish augite (bottom right) and olivine serpentinized along cracks (centre) lie in a matrix composed mainly of plagioclase laths, augite grains, and iron ore granules.

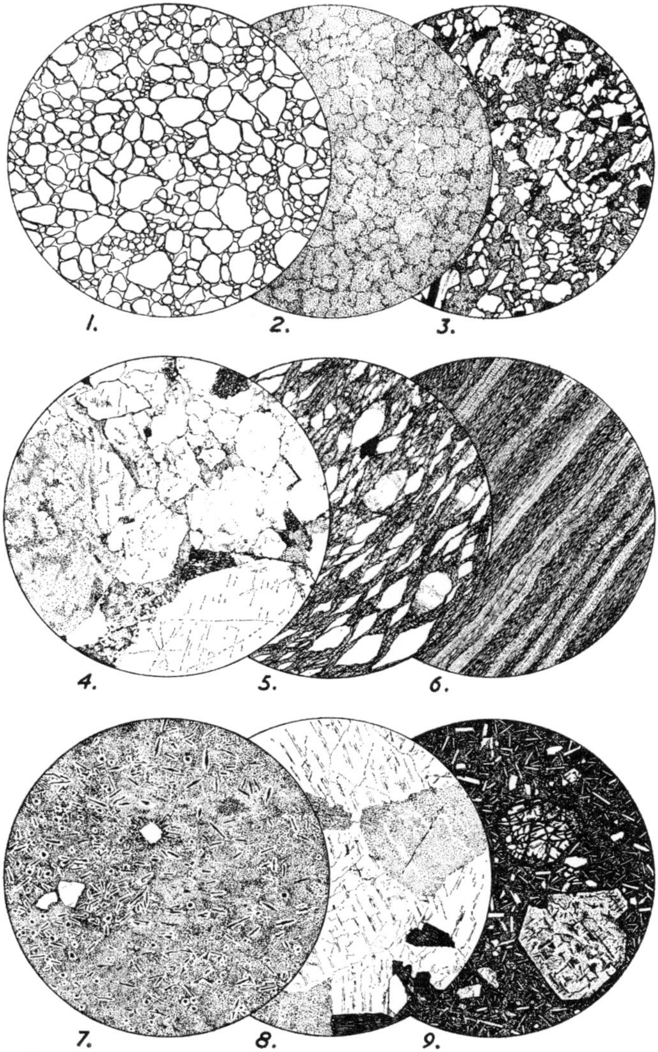

PLATE II. Micro-sections of some typical Arran rocks.

order to accommodate the amount of igneous material represented by the swarm of dykes, considerable stretching of the earth's crust has taken place. Tyrrell (1928, p. 249) has shown that in Arran over a belt of country 23·8 km wide, containing 525 measured dykes with a combined thickness of 1,650 m the stretching of the earth's crust in an ENE direction, i.e. perpendicular to the most common dyke direction, is 1 km in 14·4 km.

7c. Later Tertiary Denudation

At the close of the volcanic episode the land area of the western margin of Scotland was probably much more extensive than it is today with much of what is now continental shelf above sea level. The youngest rocks exposed at the surface were the immense piles of lava (plateau basalts) punctuated at intervals by major central volcanoes the deeply eroded remains of which are now seen in the intrusive complexes of St. Kilda, Skye, Rum, Mull, Ardnamurchan, Arran, and the now totally submerged centre at Blackstones Bank, to the west of Islay. This land surface possibly extended to the south to link up with the basalt plateaux of Antrim in the north of Ireland. The rest of Scotland was probably an undulating tableland floored largely by remnants of Cretaceous and older Mesozoic sediments and diversified by upland tracts composed of more ancient formations. It was out of a continental terrain of this nature that the existing physical features were slowly evolved throughout a long interval of time extending to the beginning of the Glacial period. Impressive evidence of its duration can be obtained by a rough estimate of the amount of denudation involved. In Arran, for example, the volcano which rose on what is now the Central Ring Complex has been cut down to its roots. The Northern Granite, a coarse-grained plutonic rock which crystallised slowly under a cover of country rock hundreds of metres thick has been laid bare to erosion. This cover doubtless included representatives of most of the formations present in the island, from Dalradian to Cretaceous. Not only has all this cover been worn away in the course of time and the underlying granite exposed but the latter itself has been subjected to prolonged erosion. "There may have been as great a thickness of granite worn away from above the ridge of Caisteal Abhail and Ceum na Caillich as has been cleared out of the valley below" (Geikie 1910, p. 25).

The stages by which this process of denudation was accomplished must be largely a matter for conjecture. It may be inferred that there were periodic earth movements producing a succession of topographies and setting in train successive fresh cycles of erosion. There was certainly at one period a series of powerful crustal disturbances which locally folded the lavas and underlying rocks and renewed movement along some of the old NE–SW lines of fracture. George (1960, pp. 98–99) has drawn attention to the striking effects of these earth movements in Antrim and has referred them to a mid-Tertiary "Alpine" period of deformation. The suggestion may also be hazarded that tensional stresses in the earth's crust related to the early opening of the northern part of the North Atlantic may have contributed to the breaking up of the western volcanic land as the result of differential or block subsidences which in places carried the Tertiary lavas and underlying Cretaceous sediments below present-day sea level. Thus the denudation which took place in Tertiary times was acting on what was already an irregular surface hence the difficulty that has been experienced in reconstructing the history of the modern landscape. For a general account of the Tertiary and Quaternary evolution of Scottish landforms the reader should consult Sissons (1976).

The geomorphology of Arran has been described by a number of writers, notably Smith (1896), Gunn (1903), Mort (1911, 1914), Scott (1918), Gregory (1920) and Tyrrell (1928). Tyrrell's account, which closely follows the views of Gregory, includes a sequence of events that can be summarised as follows:

(i) Production of an elevated undulating plateau surface, 600 m to 900 m above sea level:

(ii) Development of consequent rivers:

(iii) Uplift, with development of tectonic valleys to east and west isolating Arran from the mainland:

(iv) Prolonged denudation, with the formation of a low plain of marine erosion.

(v) Uplift with the formation of the 300 m Platform:

(vi) Prolonged denudation, with enlargement of the valleys to east and west.

The validity of the above sequence depends very much on how the physiographic features of the island are interpreted, in particular the nature of the valleys and the recognition of platforms.

(1) **Platforms.** As already mentioned (p. 17) the 300 m (1,000 foot) Platform is a striking feature in the landscape. It forms a part of the Intermediate Platform of Peach and Horne (1930, p. 2), a feature widely represented in the Scottish mountains. It makes a well defined shelf round the Northern Granite, with a maximum width of about 800 m, and slopes gently upwards to the granite mountains. Its height varies from a little below to a little above 300 m. Following the views of Tyrrell and his predecessors George (1966) emphasised the view that this, along with other erosion surfaces which he recognised in the Hebrides, was the result of marine abrasion but the great width of the 300 m surface places some difficulty on this interpretation.

There is evidence that Scotland was experiencing a subtropical climate during the late Tertiary era. In areas of the world where such conditions prevail at the present day one of the dominant land forms is the wide plain of subaerial erosion, often associated with the steep-sided mountains known as inselbergs. It requires little imagination to visualise the northern peaks of Arran as inselbergs rising above an erosion surface the remains of which are preserved as the 300 m platform.

(2) **Valleys.** The through-valley which cuts across the island from Loch Ranza to Sannox is regarded as a relic of the early consequent drainage. This depression now holds the northwest flowing Chalmadale and the southeast flowing North Sannox Water, with a watershed at about the 200 m level. Scott (1918, p.

93) suggested "that the depression following the line of Glen Rosa and passing west of the Clauchland Hills to the channel between Holy Island and Kingscross may have also been part of the consequent drainage". Of the three major valleys cutting across the granite from north to south the same writer, following a suggestion made by Smith, regarded Glen Iorsa and Glen Rosa as perhaps developed along "synclinal" axes on the undulating surface of the granite, while the Glen Catacol–Glen Scaftigill depression lies along a shatter-belt. In the northern part of the island the minor streams are in the main radial to the granite mass and some have cut gorges in the 300 m Platform. In southern Arran the drainage pattern would appear to depend more closely upon geological structure. Some of the streams, like the Kilmory and Sliddery Waters follow the dip slope of the New Red Sandstone in a southwesterly direction; the courses of the others, such as Glen Craigag and the upper part of Glen Ashdale have been guided by northwest–southeast dykes; others again are radial to the dome-shaped area of the Central Ring Complex.

The Arran valleys show many of the features characteristic of glacial action—steepened walls, flat floors, truncated spurs, etc. According to Scott and Gregory, however, there is little evidence of overdeepening, indeed this concept is coming into increasing disfavour as knowledge of the mechanisms of glacial erosion increases (see Price 1973, p. 53). It seems more likely that lateral widening of the trunk valley produces hanging valleys.

Differences of opinion have been expressed as regards the origin of the deep valleys to the west and east of Arran, which isolate it from the mainland. Scott suggested that they originated as parts of the consequent drainage. He pointed out that one of the consequent rivers may have continued from lower Loch Fyne along what is now the Sound of Bute, while another developed along the North Channel. "It seems feasible", he added, "that part of the Firth of Clyde, at least, is due to tributaries of the latter cutting back and capturing the head waters of the more northerly stream. Kilbrannan Sound may be accounted for in this way." Gregory, however, assigned these valleys to tectonic movements (which he also regarded as having been responsible for the origin of the north–south valleys in the Northern Granite). "The general course of the eastern and western coasts has been determined by faults and fractures. Kilbrannan Sound to the west and the main Firth of Clyde to the east are probably both sunken blocks, which have been let down by north and south trough-faults; and the block of North Arran was doubtless rent by north and south clefts."

While Gregory's views on the valleys in the Northern Granite may not be fully acceptable his suggestions regarding the origin of the Arran coastlines have been vindicated, at least in part, by marine geophysical surveys carried out in the Firth of Clyde (McLean and Deegan 1978, fig. 8·1) which have revealed major faults parallel to the north and eastern coastlines of Arran, and deep sedimentary basins around most of the island. There can be little doubt, however, that the final shaping of the outlines of the island have to be attributed to the erosive events of the Quaternary Period.

8. Quaternary Period

The broad divisions of the Quaternary Period are (1) Pleistocene or Glacial Epoch and (2) Recent or Postglacial Epoch. A further subdivision of the Glacial Epoch into (*a*) time of widespread ice-sheets and (*b*) Lateglacial or time of retreat of the ice has been adopted here (see table on p. 68).

8a. Time of widespread ice-sheets

This interval, often referred to as the time of maximum glaciation, began about two million years ago with the onset of arctic conditions increasing in severity until almost the whole of Scotland was covered by a mantle of slowly moving ice reaching thicknesses of hundreds of metres. It is with the effects of glaciation on a land surface of varied relief that we now have to deal. Its effects are strikingly displayed in the abraded, scoured and striated surfaces over which the ice, with its embedded rock fragments has passed and in the vast quantities of debris torn from rock surfaces on its passage and carried forward to be finally deposited as ground moraine.

Detailed studies of these effects have shown that they cannot be ascribed to one widespread advance and retreat of the ice. It is clear that one general glaciation cannot account for the diversity of direction of ice-flow. Successive major trends of ice movement on a regional scale have been recognised in Caithness, in Aberdeenshire, in Kincardineshire, and in the Midland Valley, where they have left behind them successive deposits of boulder clay differing in character and content and sometimes separated by bands of stratified sands and gravels. The cause of these successive major ice flows must be looked for in the variations in pressure exerted at different times by Scandinavian ice off the east coast and by ice radiating from the Highlands and Southern Uplands.

Sometimes the deposits left behind by an ice sheet show that for part of its course it has travelled over the sea bed. Thus, the ground moraine of an early ice sheet crossing Aberdeenshire and Kincardineshire from the northwest brought with it Arctic marine shells dredged from the Moray Firth. Another ice flow moving from the West Highlands down the Firth of Clyde was deflected eastwards, perhaps by Irish Sea ice, and carried inland over central Ayrshire clays containing Arctic shells such as *Pecten islandicus, Cyprina islandica* and *Astarte borealis*. To the same southward movement of West Highland ice may be attributed the occurrences of Arctic shell beds present in the lower portions of the Kilmory and Sliddery waters on the south coast of Arran at heights of between 21 m and 55 m. These beds were described by Watson in 1864 and by Bryce in 1865 and 1872, while the accounts of these writers have been summarised by Tyrrell in the Arran Memoir of 1928 (pp. 262–263). Bryce expressed the view that the shell beds were laid down during a period of depression of the land by at least 55 m and that the overlying "drift" deposits were formed partly by local glaciers, partly by streams. It seems preferable on the whole, however, to assign them to land ice, as in the central Ayrshire examples.

8b. Lateglacial Interval or Time of Retreat

In the closing stages of Glacial times all the conditions that marked its beginning took place in reverse order. The great ice-sheet gradually dwindled and broke up, giving way to a confluent glacier stage and then to a valley glacier stage, finally disappearing as the climate grew milder and milder. The retreat was also accompanied by changes in sea level of which a brief account is given later.

(1) **Confluent Glacier Stage.** This may be said to mark the first stage in the dissolution of the ice. There was still a general covering of ice over the country fed by glaciers radiating from a number of centres of dispersion. Many of the effects of erosion and sculpturing by ice must be set down to this stage. At this time ice was probably moving slowly southwards from the Argyllshire highlands down Kilbrannan Sound and the Firth of Clyde, sweeping across Arran where the Goatfell massif stood up as a nunatak dividing the flow into two streams.

The vast quantity of broken and ground-up rock fragments embedded in the lower layers of the ice were deposited during its retreat and left behind as an irregular hummocky mantle of rock debris in a clayey or sandy matrix. Examples of this type of deposit, generally known as boulder clay, can be well seen in some of the streams in south Arran, e.g. in Glen Ashdale, in the Sliddery Water, etc. where its usually reddish sandy character shows that it is in a large measure of local origin and derived from the New Red Sandstone of the district. Much of the lower moorland country in Arran has a cover of stony ground-moraine material with a hummocky and occasionally drumlin type of topography, often referred to under the general term of morainic drift.

(2) **Valley Glacier Stage.** In the course of time the ice was more and more confined to the major valleys. The glaciers occupying these valleys were still powerful agents of erosion, continuing the process of sculpturing their profiles. They also left striking

examples of ice-borne materials in the form of lateral and terminal moraines marking stages in their retreat. There are many good examples of moraines in Arran: for instance, in Glen Rosa and Coire a'Bhradain (Excursion 1, locs. 10 and 13); in Coire Lan at the head of the "White Water"; in Coire nan Larach (Excursion 5, loc. 9a); at the head of Glen Sannox and elsewhere. In Glen Dubh at the head of Glen Cloy there is a series of concentric moraines fronting an alluvial flat which must at one time have held a small temporary glacial loch. Sometimes the moraines take the form of an irregular assemblage of small mounds.

(3) **Corrie Glacier Stage.** Ice erosion at valley heads often led to the excavation of the amphitheatre-like hollows known as cirques or corries. These recesses, into which the ice finally shrank, were long occupied by small corrie glaciers representing the last phase of active ice. A good example in Arran is Coire Fhionn Lochan (Excursion 10, loc. 26).

The corrie and valley glacier stages described above were probably repeated during the Zone III Loch Lomond Readvance (see Table on page 68; and Price 1982, p. 67), during which time ice was quite extensive in the northern granite hills of Arran.

(4) **Fluvio-glacial Deposits.** Under this heading come deposits of gravel, sand and silt carried by melt-waters from the front of the receding ice. This drainage escaped in a number of temporary and shifting streams and the materials borne away by these were spread out irregularly to form what is known as an outwash fan. Much of the moundy deposits of sand and gravel in the lower valley of the Machrie Water and some of the spreads of gravelly debris flooring the valleys outside the moraines represent outwash material.

Other fluvio-glacial deposits were formed around (and sometimes over) bodies of dead or stagnant ice which had become cut off from the main body. There is a diversity of features of such types to be seen from the road between Shiskine and Glenloig Farm. These include kame terraces, kames and eskers.

(5) **Changes in sea level.** The later stages of the Pleistocene Epoch were accompanied by changes in sea level which have left marked features on the coastlines over wide areas. These changes, indicating alternate emergence and submergence of the land, may be correlated with the growth and decay of the great ice-sheet. Two factors involved in both growth and decay are: (*a*) the withdrawal from the sea of vast quantities of water necessary to form the ice-sheet and its release during the retreat phase: (*b*) the depression of the land under the immense weight of the ice-sheet and its slow, intermittent recovery with the disappearance of the ice.

At the beginning of Glacial times sea level stood very considerably lower than it does today, probably by at least 90 or 120 metres. Towards the close, however, a slow submergence took place and the sea rose on the land in the wake of the retreating ice. It was during this phase that the terrace feature known as the High Lateglacial Raised Beach (formerly called the 100 foot Raised Beach) was formed. The inner margin of this terrace is seldom defined by any definite erosional features. Although it forms a well-marked terrace at many localities on the Scottish coast its actual height above present-day sea level varies somewhat from place to place.

Donner (1959, 1963) and Sissons (1963, 1966 and 1967) have shown that there is a series of Lateglacial shorelines in Scotland at altitudes varying from 40 to 12 metres. Each of these shorelines is tilted towards the Central Highlands, which was the main centre for ice dispersion in Scotland. As the ice was thickest in this area, the depression of the land under the weight of the ice was greater here than elsewhere and consequently it is in this area that the greatest amound of compensatory uplift has taken place. The result is that features produced at sea level when the ice melted may have now been lifted many metres above sea level, the amount of uplift varying with distance from the former ice-centre.

That the High Lateglacial Raised Beach was contemporaneous with the valley glaciers is shown by the fact that in many cases its deposits merge into or underlie valley glacier materials, while its absence in a number of the sea lochs along our western coast is ascribed to the presence of glaciers still active at that time. In Arran valley glaciers probably excluded the Lateglacial sea from all but the lower parts of Glen Rosa and Loch Ranza but in the southwestern part of the island where the lower Iorsa was a sea-loch the sea penetrated up the Machrie and Black Waters to flood a large part of the Shiskine plain. The coastal hill ridge which includes Torr Righ Beag (107 m), Torr Righ Mòr (136 m) and Drumadoon Fort (about 60 m) must then have been isolated. The deposits of the Lateglacial sea, where exposed in Glen Rosa, in the Iorsa near Dougrie Lodge, and elsewhere, are all coastal sands and gravels. In the wider estuary of the Clyde, however, deposits of the same age include deeper water offshore clays containing marine shells of sub-arctic or arctic types. Following upon High Lateglacial Raised Beach times there was a gradual emergence of the land or retreat of the sea. That there were halts in this uplift is shown by the occurrence at various localities round the Scottish coasts of a beach feature at about 15 m, or sometimes a group of these at heights between 14 and 23 metres. There is evidence in some of the western sea-lochs that glacial activity had not ceased at the time of the formation of the 15 m terrace feature. The possible interpretation of the terraces at this level in some areas as fluvio-glacial outwash fans still leaves them within the Lateglacial period of retreat.

In Arran, Gunn (1903, p. 139; and in Tyrrell 1928, p. 265) has drawn attention to "the evidence of several beaches at different levels rising from 40 to 50 feet (12 or 15 metres) to over 100 feet (30 metres) above the sea, which are beautifully exhibited near the mouths of the Iorsa and Machrie streams". He also pointed out that these marine terraces "have evidently suffered much denudation by the action of the sea along the exposed part of the coast, and by streams at their mouths, and in the bays of Brodick and Lamlash, where several burns enter the sea".

Lateglacial times came to an end about 11,000 years ago. In the interval between it and the succeeding Postglacial epoch there was a marked amelioration in climatic conditions.

8c. Postglacial Times

Under this heading are included the various geographic and climatic changes that followed upon the final disappearance of the ice. The former are briefly described on paragraphs (1) to (3) below, the climatic changes in paragraph (4). A tabulate statement showing the succession of events is given on page 68.

(1) After the formation of the terrace feature at or about the 15 m level emergence of the land continued until sea level stood somewhat lower than it does today. The evidence for this lies in the occurrence of submerged peat-beds, remains of forest trees and old land surfaces at a number of widely spaced localities round our coasts. For example, in excavations and borings in the estuaries of the Clyde, Forth and Tay, a submerged peat-bed has frequently been recorded lying between the arctic clays of the High Lateglacial Raised Beach and the deposits of the later Main Postglacial Shoreline (formerly called the 25 Foot Raised Beach) or carse clays. The peat-bed contains nuts and pieces of bark of both hazel and birch. This stage represents a halt of sufficient duration to allow the growth of trees and, later, of their destruction by the invasion of peat mosses. It corresponds to the Boreal phase of the climatic succession.

(2) **Main Postglacial Shoreline.** This marine terrace, formed during the submergence of the land which followed Boreal times, is a striking feature round many parts of the Scottish coast. Its wave-cut platform backed by cliffs with old sea-caves, and often showing old stacks, distinguish it from any of the earlier beaches. It is well developed in Arran, notably between Brodick and Sannox, between Loch Ranza and Catacol and north of Drumadoon. Its deposits carry a littoral marine fauna similar to that of the present day. A good section of the shelly sandy clays of this beach was exposed during excavations at the site of the Post Office building in Brodick.

It may be noted here that the height of the shoreline varies slightly from place to place. In the Forth Estuary, for example, its inner margin rises gradually from a level of 7·5 metres at Bo'ness to about 14 metres near Stirling. Skeletons of seals and stranded whales have been recorded from the deposits of this beach. The period of its formation corresponds with the Atlantic phase of the climatic succession.

(3) After the formation of the Main Postglacial Shoreline there followed a fall in sea level and the final establishment of present-day conditions. There is evidence in parts of the coastline of northern Scotland of a halt in this process and the formation of a terrace at the 4·5 metre level. There is also some evidence to suggest that the latest movement has been a slight submergence which may still be in progress (Callander 1929).

(4) **Climatic Changes.** The succession of events outlined in the preceding paragraphs was accompanied by corresponding changes in the climatic conditions from sub-arctic to temperate and warm. These changes were linked in turn with variations in the type of forest trees as revealed mainly by pollen analyses. In pre-Boreal times the climate was cold and wet, with a limited forest growth of willow, birch and pine. The increased land area of Boreal times brought a climate of continental type, dry and warm summers and cold winters, accompanied by the spread of such trees as the oak, elm and alder. At this time Britain was still part of the continent with the North Sea and the English Channel largely fenland and the Baltic a fresh-water lake (Lake Ancylus). It was during the Atlantic phase which followed that the term insular could for the first time be rightly applied to Britain when the Straits of Dover were finally breached and the North Sea came into existence. Britain was now an island with a mild, wet climate, probably milder and wetter than it is today. Mixed deciduous forests covered much of the lowlands and such northern trees as the pine and birch were confined to the higher ground.

The sequence of geological and climatic changes outlined in the preceding paragraphs are set out in tabular form on page 68. The data used in compiling this are only in part geological. Much additional evidence is now available from the work of botanists using the technique of pollen analysis to provide a vegetational succession. The pollen zones recognized are shown in the extreme right-hand column of the table. Still more recently the method of radiocarbon dating has been applied to the problem of Quaternary chronology with the promise of an exactitude hitherto unknown. A brief explanation of the principles underlying the method of radiocarbon dating is given by Godwin (1961, pp. 288–289).

GEOLOGICAL EVENTS		CLIMATIC PHASES	APPROXIMATE TIME-RANGE B.C.	POLLEN ZONES
Period of Maximum Glaciation	....	....	Ending at 23,000	
Lateglacial Period:[1]				
Confluent Glacier Stage		Arctic		
Valley Glacier Stage	High Lateglacial Raised Beach	to	23,000 to 9,000	I to III
Corrie Glacier Stage		Sub-Arctic		
Final Disappearance of the Ice	....	....	8,300	
Postglacial Period:				
Submerged Peat and Forest Bed		Pre-Boreal Phase	8,300 to 7,000	IV–V
		Boreal Phase[2]	7,000 to 5,500	VI
	Main Postglacial Shoreline	Atlantic Phase	5,500 to 3,000	VII
		Sub-Boreal Phase	3,000 to 1,000	VII
Present-day Conditions		Sub-Atlantic Phase	1,000 to 0	VIII

[1]During this period there were two re-advances of the glacier ice, their limits marked by prominent moraines: these are the Perth and Loch Lomond Re-advances. The latter took place after an interval between 14,000 and 11,000 years ago when ice disappeared from Arran only to be followed by a deterioration in climate again before the final retreat of the ice at about 8,300 years ago.

[2]Material from a bed of peat underlying the carse clays of the Main Postglacial Shoreline (25 foot Raised Beach) at Airth Colliery, southeast of Stirling has been assigned by carbon dating to 6,461±157 B.C. (Godwin 1961).

TABLE 2. Lateglacial and Postglacial events.

3 ARCHAEOLOGICAL NOTES

The Boreal and Atlantic phases of Table 2 (page 68) fall within the Mesolithic or Middle Stone Age of the Archaeological records. The first definite evidence of the presence of prehistoric man in Central Scotland may be assigned to the latter part of the Boreal phase. At a site near Perth a canoe of Scots pine, hollowed out by fire, was discovered under three metres of carse clays—a relic of Mesolithic fisher folk who used harpoons of bone. The evidence for early man during the Atlantic phase takes the form mainly of heaps of discarded shells of edible molluscs, known as "kitchen-middens". These are found at various localities along the south shore of the Forth estuary, on Inchkeith, and elsewhere, resting on the Main Postglacial Shoreline. The people who left them were nomadic hunter-fisher folk, using implements of stone, bone, and deer antlers and migrating in small bands from one temporary settlement to another.

There is no extant evidence in Arran for the presence of these people, but a large kitchen-midden occurring at Glecknabae on the western shores of the island of Bute suggests that they may well have reached Arran also. So far as Arran is concerned, however, the earliest inhabitants who have left clear evidence of their presence were the Neolithic or New Stone Age people who constructed the great chambered, megalithic burial cairns distributed mainly throughout the southern part of the island. Our story now leaves the domain of geology for that of pre-history, but one point should first be emphasised. If one of these early inhabitants were to revisit the island today he would find that it showed the same broad topographic features as it did in his time. He would remember his old Arran as a land of widespread peat-mosses and marshes, with extensive woodland and scrub areas where the wild animals he hunted—deer, wild ox, wolf, etc.—found shelter. In the intervening centuries continued frost action has splintered and rent the high tops and rock screes and landslips have altered their shapes. Large quantities of rock debris have been washed down the gullies on to the valley floors and carried finally to the sea. Here and there the coastline has been slightly altered. Yet despite all the many changes that nature has wrought the essential features in the configuration of the island remain the same.

Neolithic age: (*circa* 2500–1750 B.C.). Some seventeen megalithic cairns, in a more or less ruinous condition, are still extant in Arran. A description of fourteen of the cairns has been given by Bryce in *The Book of Arran* published by the Arran Society of Glasgow in 1910. The essential elements in these structures were a forecourt and a portal or entrance leading into an elongated chamber divided transversely into compartments for the dead. Portal and compartments alike were lined and roofed with stone flags and the whole structure covered by stones and earth. Sometimes two or even three chambers were present. The cairns also had an edging of incurved stones round their margins. Good examples are the Carn Bàn cairn at the head of the Kilmory Water, the East Bennan cairn and the cairn at Whiting Bay known as the "Giants' Graves", all three with a length of about 30 m and a width of about 18 m. All the implements obtained at the various sites were of stone—knives, acrapers and arrowheads of flint, polished stone hammers and axes, and at least three arrowheads of Corriegills pitchstone. Pottery remains consisted of rough bowl-shaped vessels made of dark clay, with a rounded base and either unornamented or simply ornamented with lines.

The people who built these cairns, or long cists as they have been termed, were immigrants from the south who appear to have made their way northwards via the Irish Sea and North Channel to establish small settlements in Northern Ireland and along the southwestern Scottish coast. They brought with them some knowledge of agriculture and of domestic animals (bones of ox, pig and sheep have been found in the cairns); they could fashion rough hand-moulded vessels of clay; they were skilled makers of stone implements and weapons and must have had some acquaintance with carpentry. A clue regarding the dates at which they lived in Arran has been provided by the Monamore chambered cairn, southwest of Lamlash. Re-excavation in 1961 yielded charcoal from a small fire lit in the forecourt area. This has been dated at 3160±110 B.C. by C^{14} and another similar hearth, lit not long before the tomb ceased to be used, was dated at 2240±110 B.C. The Neolithic cairn builders thus seem to have reached Arran before 3000 B.C. in radiocarbon years, which is probably nearer 4000 B.C. in real years according to the tree-ring calibration of C^{14} dates.

Early Bronze age. The builders of the chambered cairns were followed by a fresh influx of invaders belonging to a different race, who brought with them their own customs and traditions and their own material culture. These newcomers were members of a mainly nomadic people, the Beaker folk (so named from the type of pottery associated with them—a beaker or drinking vessel with a tall, straight-sided neck and a globular base), who made their adventuresome way across the North Sea and English Channel to occupy much of Britain, penetrating in course of time as far north as Caithness and the Orkney Islands. They introduced the practice of burying their dead singly in what are known as short cists, consisting typically of four rock slabs set up vertically to enclose a space roughly 1 m long, 0·6 m wide and 0·6 m deep and covered by a heavy stone lid. There are a number of such short cists in Arran. Sometimes they occur in the middle of rounded tumuli or cairns built of earth and stones, their outer margins defined by an edging of large rock slabs. Examples are the cairn known as "Ossian's Mound" at Clachaig (10·6 m by 8·2 m in diameter) and the cairn at Brown Head (7·9 m in diameter). Sometimes the cists occur in the centre of the areas marked out by the great stone circles of Machrie Moor and other localities. Often, however, there is no structure above ground to mark their site. While still using implements of flint (knives, scrapers and arrow-heads) and stone (hammers and axes) these people brought with them a knowledge of bronze. An ornamental bronze dagger, 23 cm long and 7·6 cm wide at the base, was recovered from a short cist interment at

South Feorline, Blackwaterfoot, together with a gold band or fillet ornamented with parallel flutings; while a bronze pin 64mm long, probably part of a brooch, was found in a cist in the centre of one of the stone circles on Machrie Moor. At a number of localities oval beads, perforated discs, and plates of jet were recovered, representing portions of necklaces. All the pottery remains found in the short cists in Arran are of the "food-vessel" type, flat-bottomed urns, 12 to 15cm high with circular raised mouldings and an irregular, often rich pattern of ornamentation. The "food-vessel" is a somewhat later type of urn than the beaker type. The occurrence of bronze weapons and of bronze, gold and jet ornaments implies the development of trade routes and some system of commerce. The bronze and gold probably came from Ireland, while the jet, a compact, black resinous material derived mainly from fossilised coniferous wood, originated near Whitby in North Yorkshire, England, where nodules strewn on the shore are derived from a 9m bed of Upper Lias shale in which the jet occurs as irregular lenticular masses.

Further evidence of the age of the Beaker folk in Arran was found in 1959 when a short stone cist grave was discovered on Broombrae Farm near Blackwaterfoot. It had apparently contained a crouched skeleton but most of the bones had disintegrated. Also in the cist were three flakes of flint and pitchstone and a Food Vessel. The latter dates the grave in the early Bronze Age, approximately in the middle of the second millennium B.C.

While the late-Neolithic and early Bronze Age cultures were in many ways distinctive, they possessed in common one striking characteristic, the building of megalithic (or big stone) structures. The great stone tombs of the Neolithic people (the long cairns), and the round cairns, the monumental stone circles and monoliths of the early Bronze Age folk would seem to represent successive phases of a transitional and compelling urge to honour their dead. In this respect the two cultures may be regarded as continuous. There is also evidence of a period of transition during which there was a blending of the two. The people of the chambered cairns appear to have gradually adopted some of the burial customs of the round cairn folk. The transition is exemplified in the Dunan Beag and Dunan Mòr cairns above Lamlash (Excursion 3, loc. 10). Here and at other localities there is a marked simplification of the structural features of the chambered cairns, associated with the presence of ornaments and pottery characteristic of the short-cist culture—jet beads and urns of both beaker and food-vessel types.

The houses of these early peoples were probably free-standing circular structures with timber frames. Some of the "hut circles" around Machrie Moor date from about this time although others may be later, having perhaps been built during the Iron Age. Of the language spoken by the Beaker folk we have no knowledge.

There follows now a long gap, covering many centuries, in the story of Arran. Doubtless also there was a slow advance in culture with improvements in methods of metal-working, in the manufacture of implements and weapons, in husbandry and in pottery-making. Perhaps weaving, not known in Britain before the Bronze Age, may be added. But there are no remains to guide us in these "dark ages". We do not know when the first Gaelic speakers colonised the country. These were Celtic peoples, driven by population pressures from the Continent, who began to invade Britain, probably about 1000 B.C. and in increasing numbers from about 750 B.C. It was these Celts who introduced the art of iron forging. In course of time they spread northwards to occupy and dominate Scotland and Ireland. When this happened we can only surmise. There would appear, however, to be no certain evidence of the use of iron in Scotland much before the first century B.C. Up to that time the inhabitants were still in the Bronze Age.

To conclude this synopsis reference must be made to two events which bring the story of Arran down to well within historical times. These are (i) the establishment of religious centres by missionaries from Northern Ireland, and (ii) the Norse Settlements.

(i) During the latter part of the 5th century and the first decades of the 6th century colonists from Ireland had settled in considerable numbers in western Scotland and about A.D. 575 they founded the loosely-knit kingdom of Dalriada, which included much of

Argyllshire and the neighbouring islands. Missionaries from Irish monasteries accompanied and followed their countrymen to establish outposts of the Irish church and several of these visited and lived for a time in Arran. St. Brendan, whose voyages among the western isles in search of the earthly paradise was a favourite theme in medieval legends, gave his name to Kilbrannan Sound and would seem to have founded a small cashel or church settlement in the hillside above Kilpatrick, probably about the middle of the 6th century. Little but the circular foundations can now be seen. About A.D. 680 St. Molios settled on Holy Island, Eilean Molaisi, as it was formerly called (Excursion 11b).

(ii) The period of the Norse invasions and settlements in Scotland lasted from about A.D. 800 to 1263, the date of the Battle of Largs. In Arran the Norse invaders have left evidence of their presence in the form of two burial cairns. One of these, at Kingscross Point, yielded, in addition to some calcined human bones, nails and rivets such as were used in the making of the Viking galleys, pieces of bronze and a bronze coin. The last-mentioned item placed the date of the burial as probably in the latter half of the 9th century. The second burial site which lies on the south side of the Blairmore Burn, above Lamlash, yielded the remains of an iron sword and shield resembling types found in graves in Norway dating from the later 8th or early 9th centuries. Many of the place-names also bear witness to the Norse occupation and are noted in the glossary.

4 DESCRIPTIVE ITINERARY

The fourteen excursions described in this section of the Guide are planned to serve as an introduction to the salient features of the geology of Arran. They vary in length and range of subject matter covered. Each excursion is illustrated by a map upon which the various exposures to be examined have been indicated by a number. Excursion 1 is the widest in its scope, introducing the beginner to examples of sedimentary, igneous and metamorphic rocks of a varied character and of diverse ages, Excursions 2 and 3 cover the ground immediately south of Brodick where sandstones and breccias of Permian age and the numerous intrusions cutting them are magnificently displayed. Excursions 4 and 5 deal in the main with the Carboniferous System and Excursion 6 with Old Red Sandstone and Ordovician rocks. Excursions 7 to 10 are devoted mainly to an examination of the metamorphic rocks (Dalradian schists etc.) on the northwest and west margins of the Northern Granite, and include (Excursion 7) a visit to the classic unconformity at North Newton. Excursions 11 to 13 are concerned with a study of the geological features characteristic of the southern half of the island. Excursion 14 is the most detailed and, in part at least, the most specialised, since it covers an area of great complexity of rock types and structures forming the northwest portion of the Central Ring Complex.

Suggestions for shorter itineraries: For those who have their own transport but a very limited time available a broad sample of the geology of the island is provided by the Road Logs in Section 5 of this Guide (p. 179). Otherwise the following abbreviated itineraries are recommended to visitors who wish to sample a wide range of geology but do not want to follow all the itineraries:

(*a*) *Nine-day itinerary:*

First day: Glen Rosa and Glen Shurig, Excursion 1.
Second day: Corrygills and the Clauchland Hills, Excursions 2 and 3.
Third day: Corrie and the Locherim Burn, Excursions 4 and 5.
Fourth day: Glen Sannox and North Glen Sannox, Excursion 6.
Fifth day: Catacol, Loch Ranza, North Newton and the Cock of Arran, Excursion 8, locs. 1a to 9a, and Excursion 7, locs. 1 to 17.
Sixth day: Catacol, Loch Tanna, Pirnmill and Loch Coire an Lochan, Excursions 9 and 10.
Seventh day: Kingscross and Glen Ashdale, Excursion 11a.
Eighth day: Largybeg and Bennan, Excursion 12; and (if transport is available) Drumadoon and the Tormore shore, Excursion 14.
Ninth day: Area around Ard Bheinn, Excursion 14.

(*b*) *Six-day itinerary*

First day: As above.
Second day: Sannox and Corrie, Excursion 6, locs. 1–11, 14–17; and Excursion 4, locs. 1–14.
Third day: As fifth day above.
Fourth day: Brodick to Lamlash, Excursion 2; Whiting Bay to Glen Ashdale, Excursion 11.
Fifth day: As eighth day above.
Sixth day: As ninth day above.

(*c*) *Three-day itinerary*

First day: As for six-day itinerary.
Second day: As for six-day itinerary.
Third day: Corrygills and the Clauchland Hills, Excursions 2 and 3.

Excursion 1: **Glen Rosa and Glen Shurig**

(Figure 6, localities 1 to 23)

The main purpose of this excursion is to provide the visitor with an introduction to the geology of North Arran. It covers a wide range of formations and the localities selected for examination include sections of Dalradian, Lower Old Red Sandstone, Upper Old Red Sandstone, Carboniferous and Permian rocks, as well as sections in the Northern Granite and representative examples of some of the minor intrusions (pitchstones, felsites and dolerites) of Tertiary age. The excursion also furnishes numerous illustrations of glacial phenomena and striking views of the scenery of the North Arran mountains.

In following out this excursion from locality 1 to locality 14 the visitor is descending in the geological succession from Permian through Carboniferous and Old Red Sandstone to Dalradian, whereas in the latter part of the itinerary the sequence is an upward one from Dalradian to Permian.

1. The beds exposed between Brodick Pier and the mouth of the Cloy Water (Invercloy) belong to the lower part of the Brodick Beds (see p. 37). They consist of a series of breccias made up of more or less angular fragments of rock embedded in a red gritty matrix formed mainly of rounded grains of quartz. The fragments are almost entirely quartzite, vein-quartz or quartz-mica-schist, types of rock common among the Dalradians of the southern Grampians, and vary in size from pieces 15 to 18 cm in diameter down to pebbles less than 1 cm across. The often rounded nature of the smaller pebbles should be noted, as well as the wind-rounded quartz grains present both in the matrix and in the intercalated beds of sandstone. Occasionally pebbles whose surfaces suggest that they have been polished and faceted by wind action occur. The assemblage as a whole points to accumulation during occasional rain storms or cloudbursts in a desert climate.

The general inclination of these beds is to the south at angles of about 15°. At two localities on the shore there are oval-shaped areas, 15·5 and 46 m in diameter respectively, within which the

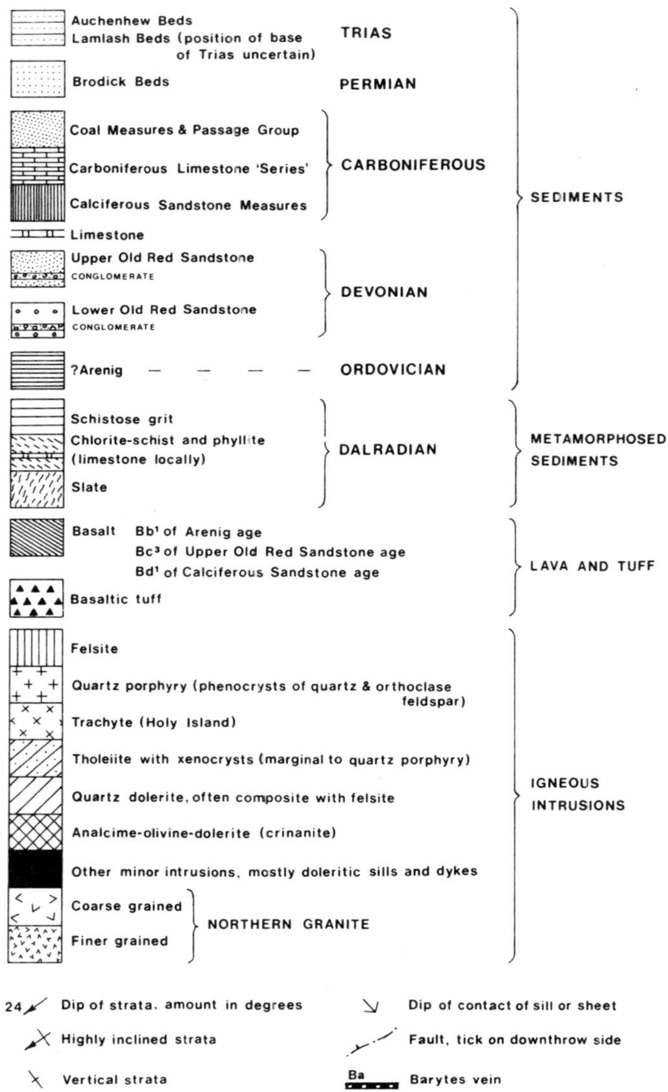

FIG. 5. Index to the geological sketch-maps, Figs. 6, 7, 9, 11 and 13–16.

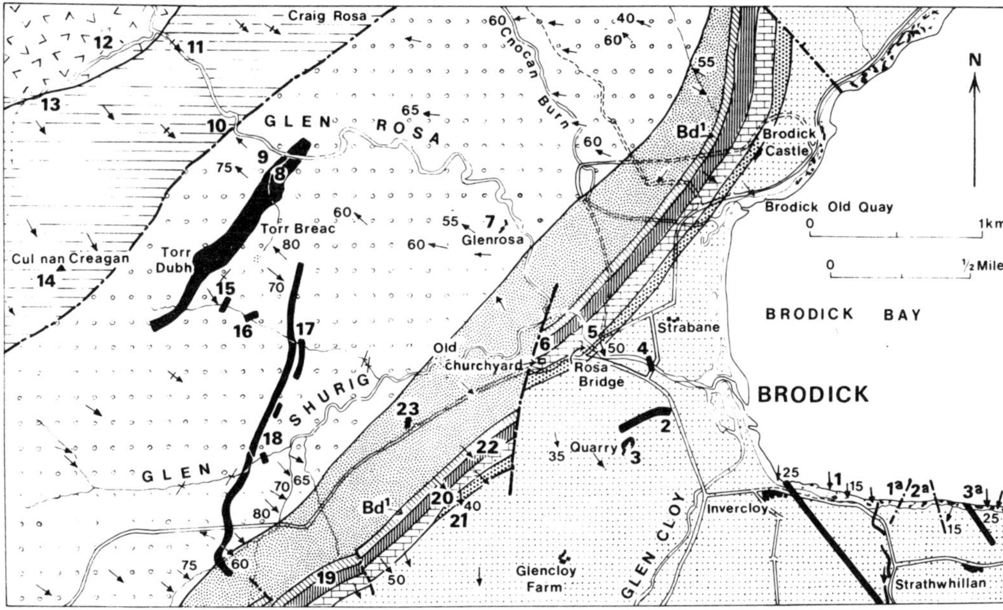

FIG. 6. Geological sketch-map of the Brodick area to illustrate Excursion 1 and part of Excursion 2.

conglomerates are much broken up and shattered so as to simulate the appearance of a volcanic agglomerate. There is no evidence, however, of the presence of any igneous material. Tyrrell (1928, p. 81) suggested that the phenomena may be due to the explosive effects of gases emitted from small, underlying intrusions of Tertiary age, but a more recent explanation (Astin and Macdonald 1983, pp. 47–58) attributes the breccias to a sedimentary origin associated with local faulting.

2. Examine the intrusion of pitchstone well exposed at the back of the schoolhouse yard and in the Stronach Wood behind (please do not hammer in the schoolyard). It can be traced for some 320 m more or less parallel to the strike of the red Permian sandstone into which it has been intruded, and the mode of occurrence suggests that it is a sill. It is a hard splintery rock, with a dark-green, almost black, colour in hand specimen and a characteristic resinous lustre. The porphyritic crystals embedded in the glassy groundmass are feldspar and olivine but some bands are almost free of phenocrysts. Specimens frequently show a striped appearance produced by differential movement of the rock (fracture flow) while still in a molten condition. A drawing of part of a thin section of the pitchstone is reproduced in Plate II, figure 7.

The schoolhouse pitchstone belongs to the same group of porphyritic Arran pitchstones which includes the Glen Shurig dyke of locality 23 and the Tormore dykes of Excursion 13.

3. Visit the old quarries in the brick-red sandstones belonging to the dune-bedded rocks at the base of the Permian succession, the Corrie Sandstone. These sandstones dip to the south-southwest at angles of between 30° and 40°. The dune-bedding and rounded quartz grains ("millet seed" grains) characteristic of the division should be carefully examined. It is often difficult to determine the true angle of dip in these rocks owing to the prevalence of the dune-bedding. Note the wide, flat tract of low ground stretching up Glen Cloy, composed in part of recent river alluvium but mainly of deposits belonging to the old raised sea-beaches. In High Late-glacial Raised Beach times (p. 63) the

Stronach ridge must have been a promontory separating the lower portions of Glen Cloy and Glen Rosa.

The standing stone at the roadside at Stronach (almost opposite the school) is a slab of red Permian sandstone measuring about 3 m in height, 1·5 m in breadth and about 0·6 m in thickness. There are a number of such solitary monoliths in Arran, the most notable being at Auchencar farm on the west-central side of the island, which rises above ground to a height of 4·8 m and may have a total length of 6·4 m. Some of the standing stones, like that at Auchencar, may have originally formed part of a stone circle similar to the well-known circles on Machrie Moor and Tormore.

4. Examine the sill exposed in the Rosa Water at the footbridge near the museum (formerly Rosaburn Smithy). It is an example of a composite intrusion, that is, an intrusion formed by successive injections of molten material of different composition. The sill, in part basalt, in part felsite, is intruded into Permian sandstones here dipping in a south-easterly direction at about 32° and the section from above downwards shows:

(1) Red sandstone.
(2) Basalt—about 1·2 m.
(3) Felsite, spherulitic and banded, with a sharp junction against (2), about 2·1 m.
(4) Felsite with fragments of basalt, about 1·2 m.

The basalt fragments in the lower part of the felsite have probably been derived from a second intrusion complementary to the first but not exposed (Tyrrell 1928, pp. 137–138 and fig. 19).

Continue along the footpath to the junction with the Brodick–Corrie road and note on the way the old channel of the Rosa Water to the west of Strabane. The mouth of the Rosa has shifted its position several times and its present course out to sea (like that of the lower part of the Cnocan Burn) is in part at least artificial.

5. The section in the Rosa Water at Rosa Bridge on the Brodick–Corrie road shows a metre or so of sandstone and shales that can be definitely assigned to the Upper Carboniferous. Their age is proved by the occurrence of thin ribs of ironstone containing

fresh-water lamellibranchs (mussels) and fragments of plant (lepidodendroid) stems. One of these ironstones, 23 cm thick, outcrops a few metres above the bridge.

Red sandstones belonging to the Permian are exposed 110 m or so downstream from the bridge, so that the junction between the Carboniferous and Permian rocks must be drawn in the gap between the two exposures.

6. From locality 5 follow the String Road for a short distance until it is joined by the branch road to Glen Rosa. Examine the exposure on both sides of the Glen Rosa road. The rocks here are sandstones belonging to the Carboniferous Limestone Series and are nearly vertical. Note the hollow feature adjacent to the northern edge of the String Road, against the old churchyard, where the Corrie Limestone (Excursion 4, loc. 8) was at one time worked on a small scale at the surface. The sandstones are composed almost wholly of angular quartz grains and some of the beds are white in colour. They closely overlie the limestone and their compact siliceous character makes them a useful index horizon in mapping the various exposures of the Lower Carboniferous succession between Corrie and Glen Shurig. These white sandstones with their angular grains should be compared with the red, round-grained sandstones of the Permian (Plate II).

Note that the outcrops are cut off to the west by a fault which shifts the Carboniferous rocks southwards to the hillside south of the String Road.

Continue along the String Road to Glen Rosa Cottages at the entrance to Glen Rosa and note how the solid rocks are nearly everywhere concealed by alluvial and glacial deposits. The pebbly red sandstones, cut by a few thin dykes, which are seen in the Shurig Water about 230 m slightly north of west of the old churchyard, belong to the Upper Red Sandstone and a small outcrop of rocks referred to the same series occurs at the roadside 310 m south of the cottages. The boundaries between the Carboniferous and the Upper Old Red Sandstone and between the latter and the Lower Old Red Sandstone can only be approximately indicated on the map.

7. Strata of Lower Old Red Sandstone age are exposed by the roadside near the Cottages and are well seen on the wooded ridge (Am Binnein) to the west. They consist of massive dull-red sandstones together with bands of purplish sandy shales dipping to the northwest or west-northwest at angles of between 50° and 60°. The sandstones, which sometimes show sun-cracked surfaces, were at one time worked for local building purposes in small quarries at the east end of the wood.

On passing into Glen Rosa note the wide U-shaped valley with the steep sides and hanging valleys characteristic of glaciated regions. The floor has been levelled by the deposition of alluvial sediments through which the Rosa follows a winding course. Good sections showing river gravels and sands can be examined at a number of places where the stream has cut into former flood-plains.

Note the roughly defined higher terrace which can be traced up the glen as far as the bend in the stream just above locality 8, a distance of about 2·4 km from the mouth of the Rosa Water. This most likely was formed as outwash deposits laid down at a time when the ice stood at the moraines near locality 10.

8. A broad, sill-like mass of igneous rock is here intruded into the Lower Old Red Sandstone sediments. It is a fine-grained grey, sometimes pinkish, rock which can be traced in a south-southwesterly direction, by Torr Breac and Torr Dubh, for about 1·6 m. Under the microscope it is seen to consist mainly of stout, lath-shaped crystals of feldspar together with pale-coloured augite in single grains or aggregates; a good deal of chlorite is also present, with scattered grains of epidote and a little quartz. Epidote also occurs here and there in veins or streaks. The rock represents a dolerite intrusion in which the feldspars have been altered and chlorite and epidote developed as new minerals.

9. Examine here the sections in hard sandstones and sandy mudstones of the Lower Old Red Sandstone, dipping steeply to the west-northwest.

10. Good sections in coarse sandstones and hardened, almost slaty mudstones, are seen in the Rosa Water close to the line along

which the Lower Old Red Sandstone comes into contact with the Dalradian schists. The position and direction of this boundary can be closely fixed and followed for some distance and its nature should be carefully studied. Anderson has recorded (1945, p. 278) that at one point, 302 m southwest of the Rosa Water, Dalradian schistose grits are in sharp contact with pebbly sandstones of the Lower Old Red Sandstone, the latter actually passing under the older rock for a metre or so. Some time should be spent in tracing and examining the line of contact here and in the little tributary where it is crossed by the Glen Rosa track and a search should be made for evidence of disruption along the junction. The variation in strike and dip of the Lower Old Red Sandstone and Dalradian should also be noted.

The Lower Old Red Sandstone sediments here have been considerably altered and are grey or greenish-grey in colour in contrast to the usual dark-red or purple hues of the series (see for example loc. 7). Epidote has developed in the gritty sandstones and imparts a greenish tinge to them, while some secondary silica seems to have formed. The purple mudstones and shales have been altered to hard, blue-grey massive rocks in which the bedding is obscured. These changes are due to contact alteration in the vicintiy of the great mass of the Northern Granite. The induration and alteration of Lower Old Red Sandstone and Dalradian rocks alike have served to obscure the nature of the junction line. This was taken by Anderson (*loc. cit.*) to represent a continuation of the Highland Boundary Fault. Tomkeieff (1961, p. 16) referring to the discordant boundary on the hillside above Glen Rosa says that "this discordance may be due either to an unconformity or a fault". Certainly the former cannot on the local evidence be ruled out (see also p. 27).

A fine series of moraines crosses Glen Rosa except where breached and cut through by the stream. They represent stages in the retreat of the glacier which occupied the glen in Lateglacial times. Good sections of the materials composing the moraines (earthy sands and gravels with boulders and stones of all sizes) can be seen in the banks of the Rosa Water a little above locality 10.

11. Here hard, bluish-coloured Dalradian schists, baked and indurated through proximity to the granite, can be examined. They show the development of contact minerals such as andalusite and cordierite, giving the rocks a spotted or speckled appearance. They are mostly altered schistose grits but some of the finer-grained bands have taken on the character of a splintery contact-hornfels. These rocks should be compared with the more typical Dalradian schists to be examined on Cul nan Creagan (loc. 14).

A short distance upstream and a few metres below the junction of the Garbh Allt with the Rosa Water is the schist–granite boundary. The actual contact can be closely located. The coarse granite near the contact is cut by veins of finer-grained granite and shows in places irregularly shaped drusy cavities in which good crystals of amethyst and smoky quartz (cairngorm) have been obtained. Stop at the footbridge over the Garbh Allt for a view looking northwards up the U-shaped valley of upper Glen Rosa with Beinn a'Chliabhain rising on the left and on the right the steep western face of Glenshant Hill, and the Goatfell massif beyond. Cir Mhòr (798 m) dominates the head of the glen and is connected to the Goatfell ridge by the low col of The Saddle (431 m) which separates Glen Rosa from Glen Sannox. Behind Cir Mhòr can be seen Ceum na Caillich, the Carlin's or Witch's Step: this deep-cut rift in the jagged outline is due to a dyke, 7·3 m wide, which has weathered away more rapidly than the granite it cuts.

Ascend the Garbh Allt, keeping to its northern bank, and note the coarse granite with its characteristic jointing, well seen where the burn rushes over the falls to join the Rosa Water.

Differing views have been expressed regarding the origin of the "hanging" junction of the Garbh Allt with the Rosa. The former flows at first southwards from Coire a'Bhradain across the 300 m ("Thousand Foot") Platform, but at about the 245 m level, where it is joined by the burn from Coire na Cuiseig, it turns east-northeastwards and cascades steeply down to the Rosa, a fall of at least 150 metres. One interpretation attributes the discordance to widening of Glen Rosa by glacial erosion. Another explanation suggests modification of the drainage pattern. On the latter interpretation the Garbh Allt was at one time the headwaters of Gleann Easbuig which joins the Machrie Water near Glenloig, but was later captured and diverted eastwards by a small tributary flowing down the west slopes of Glen Rosa. This diversion is of recent origin and has not yet been able to cut a deep channel or gorge (for references see p. 58).

12. Near here note the presence of occasional thin dykes and crush-lines in the granite.

13. At this locality there is an excellent section in the burn showing the granite-schist contact. The granite is here fine-textured for the metre or so nearest the contact. Cordierite has been recorded from the adjoining achists. The next part of the route takes the visitor southwards to Cul nan Creagan, but before leaving the Garbh Allt he should note that it rises in Coire a'Bhradain between Ben Nuis and Beinn a'Chliabhain. This corrie contains a magnificent set of moraines which are well worth a visit.

14. Good hill-top sections of gritty, sometimes pebbly schists, locally carrying quartz veins and steeply inclined to the southeast, can be examined on Cul nan Creagan (350m). From this point onwards the succession of stratified rocks encountered is an upward one, from Dalradian to Permian. From Cul nan Creagan the route lies in a direction a little south of east towards the headwaters of the Allt a'Bhrighide, a tributary of the Shurig. This burn provides splendid sections in the Lower Old Red Sandstone.

15. Note here the occurrence of fine-grained red sandstones, among which are intercalated thin beds of mudstone or shale. Some of the bedding planes show sun-cracks.

16. Examine at this locality the two dykes, 0·46 and 0·2m thick respectively, of tachylite, a black compact glassy rock of basaltic composition. Under the microscope the thinner dyke shows small crystals and needles of labradorite in an almost opaque glassy matrix.

A little farther downstream are good exposures of conglomerate composed mainly of pebbles of quartzite and quartz, with intercalated bands of fine-grained, deep-red sandstone and some purplish shales. These give way downstream to breccias in which large pebbles and fragments of andesitic lavas are numerous. This occurrence may be compared with that noted on Excursion 6 (loc. 3). No extrusive igneous rocks are known in the Lower Old Red Sandstone succession on the east side of the island. On the west side, however, an interbedded volcanic series (hornblende- and olivine-andesites) is present in a belt stretching from near Auchencar to near Garbh Thorr. The occurrence of the breccias with igneous rock fragments in the east suggests that there, too, there was a volcanic series the denudation of which provided materials to the conglomerates.

17. There is an excellent section here in a massive intrusion of felsite to which the name the Glen Shurig dyke has been given. It is about 30m wide where seen in the Allt a'Bhrighide and can be followed in a direction a little west of south across Glen Shurig and the String Road. A second 9m felsite dyke crosses the Allt a'Bhrighide just below the first. These felsites are compact, finely crystalline or cryptocrystalline rocks.

18. Follow the line of the thick felsite dyke down to its outcrop in the Shurig Water. A little downstream from this there is a good section showing sandstones with shale bands dipping southeast at angles ranging from 48° to 65°. The sandstones are highly feldspathic, purplish in colour and weather to a yellowish hue.

The dip arrows at locality 18 (fig. 6) indicate sections where fragments of fossil plants were obtained in 1897. There are two horizons at which these occur:

(*a*) in a 0·75m band of shale best exposed in the right bank of the stream, about 180m from its junction with the Allt Mòr;

(*b*) about halfway between (*a*) and the mouth of the Allt Mòr.

These plant remains are referred to *Psilophyton,* an early type of vascular plant characteristic of the Lower Old Red Sandstone of Angus, etc. It belongs to the Psilophytales, a subdivision of the Pteridophyta, one of the three non-seed-bearing divisions of the Plant Kingdom which includes the ferns, horsetails and clubmosses of the present day as well as many of the old Carboniferous swamp plants and trees. *Psilophyton* itself was a small plant possessing upright, branching, cylindrical stems some of which carried spines; the diameter of the stems varies from 1 to 7 mm. Of the Arran specimens only one could be determined specifically as *Psilophyton princeps*. This fossil plant was first discovered in Arran in the Allt Mòr, the tributary of the Shurig already referred to, some distance up from its mouth, by Sir A. Geikie and his students in 1882.

Follow the Allt Mòr upwards to the String Road, noting the stream exposures of hard, micaceous, reddish-purple and sometimes pebbly sandstones cut by thin basalt dykes. The dip steepens to about 80° near the road at the line where the upper limit of the Lower Old Red Sandstone has been drawn. Cross the road and ascend the hillside to locality 19; in doing so the outcrop of the Upper Old Red Sandstone, here concealed by drift, has been traversed.

19–22. At locality 19 and again at localities 20 and 21 to the northeast there are rocks belonging to the Carboniferous. Exposures are poor, however, and it would be advisable if the visitor were familiar with the Carboniferous succession displayed along the Corrie shore before examining the Glen Shurig sections. If so there will be little difficulty in recognising representatives of the Calciferous Sandstone lavas at localities 19 and 22, and between 20 and 22 some of the white quartzose sandstone overlying the Corrie Limestone (see Plate II, fig. 2). This limestone is not now seen here but its position is marked by a hollow along which it was quarried. All the beds have a steep dip to the southeast.

Before leaving the hillside a traverse should be made to locality 21 to see the red, cross-bedded Permian sandstones which form the crest of the ridge between Glen Shurig and Glen Cloy. The sandstones are inclined at high angles in a more southerly direction than the underlying Carboniferous. The junction between the two cannot be closely fixed.

23. At this locality, on the branch road leading to West Shurig farm, a search may be made for an interesting pitchstone dyke.

This is a dark-green rock carrying phenocrysts of quartz, pyroxene, iron-rich olivine and feldspar. These minerals occur as clusters or aggregates set in a glassy matrix or groundmass containing numerous crystallites. For a full description of this pitchstone see Scott (1915, pp. 147–150).

Excursion 2: **Brodick to Lamlash Bay by the Coast**

(Figure 6, localities 1a–3a)
(Figure 7, localities 4a–19a)

The objects of this excursion are: (1) to study the rock-types present in the two sub-divisions of the New Red Sandstone known as the Brodick Beds and the Lamlash Beds (table on p. 37); and (2) to examine some of the dykes and sills traversing the Permian along this part of the coast. Magnificent, almost continuous exposures are available in the cliffs at the back of the low raised beach platform and on the beach itself.

The great majority of the dykes along the coast here occur as vertical or nearly vertical walls of rock. In some cases, however, they are steeply inclined, usually to the northeast (see for example locality 10a). The dykes sometimes weather more rapidly than the rocks they cut and in such cases form hollow, trench-like features; sometimes they have proved more resistant to erosion than the strata around them and stand up as walls projecting above the surface; sometimes again it is the baked and hardened sediments immediately adjoining a dyke that are left projecting as two parallel ridges (see locality 5a).

1a. For some 300 m eastwards from the pier the shore has been adversely affected by industrial development of the raised beach; at high tide some of the rocks mentioned below cannot now be examined. East of the pier the shore is occupied by massive, red, cross-bedded sandstones. Note the resemblance of these sandstones, in their characteristic large-scale dune-bedding and rounded ("millet seed") quartz grains, to the Corrie Sandstone of the Corrie–Brodick shore (Excursion 4). Gunn (1903, p. 69) regarded the masses of sandstone on the shore here (see also loc. 4a) as portions of the Corrie Sandstone faulted in among the breccias, but Tyrrell and Barrett (Tyrrell 1928, p. 81) interpret them as "fossil dunes, or groups of dunes, against and around which the breccias were banked up and interbedded". Note the gashes in the conglomerates of the raised beach cliffs caused by differential erosion of the dykes.

2a. The red sandstones of locality 1a end abruptly against a north-northwesterly fault along which a dyke has been intruded: beyond this conglomerate and breccias appear in force, occupying the foreshore continuously for a distance of some 460 m. Intercalations of sandstone showing good examples of rounded quartz grains occur. The 6 m NW–SE trending dyke cutting the breccias should be examined; it is an olivine-dolerite containing laboradorite feldspar, purplish augite, olivine, some grains of iron ore and a little analcime. Note how the darker minerals (olivine, augite) are often segregated in patches giving rise, on weathered surfaces, to a characteristic spotted appearance.

3a. The shore section here shows conglomerates with many lenticular intercalations of red, "millet seed", dune-bedded sandstone. Some of the pebbles found in the sandstones exhibit wind-faceting. The main interest of this locality, however, is the occurrence in the cliffs of a considerable thickness of breccia containing large numbers of sub-angular or partly rounded boulders and pebbles of a vesicular, red-speckled, olivine-basalt lava. This basaltic horizon, with which sandstones carrying ashy material are associated, has been recorded elsewhere at the same position in the Brodick Beds and its occurrence suggests that contemporaneous volcanoes were active nearby. Some of the basalt lava pebbles resemble types common among the Permian lavas of Central Ayrshire.

The most conspicuous of the dykes on the shore is an 8·5 m wide intrusion of olivine-dolerite (crinanite type) running in a west-northwesterly direction.

This part of the shore section ends against a low-angled NE–SW directed fault which crosses it a short distance west of the little lagoon on the raised beach.

4a. It is along this part of the coast that the highest cliffs occur. Both cliffs and foreshore are formed of massive, red, cross-bedded sandstones which, it has been suggested, represent the remains of a group of dunes (c.f. locality 1a). The general inclination of the sandstones is about 15° to the south-southeast. Note the 11 m wide, NW trending dyke of olivine-dolerite cutting the sandstones about 270 m east of the little lagoon; it is inclined to the NE at about 70°.

5a. Some of the dykes here show indurated sandstones on their margins standing up as ridges. Good examples of dune-bedding can be examined in the cliffs a little to the east. Note the large erratic of coarse granite on the shore.

6a. Dune-bedded red sandstones with some intercalated bands of breccia occur along the shore, dipping to the south-southeast at angles of between 25° and 30°. Examine the broad intrusion of felsite traversing shore and cliffs in a general east-west direction. The felsite, some 24 m thick, is a light-coloured, faintly yellowish rock with a characteristic fissile appearance due to the development of joints parallel to its surface. It is inclined at about 60° to the SSE. When examined under the microscope it shows scattered microphenocrysts of quartz set in a fine-grained groundmass of intergrown quartz and orthoclase feldspar. A thin pitchstone is developed at the base of the felsite, where indurated breccia separates it from a thin tholeiitic dolerite sill.

7a. Corrygills Point. Here thin-bedded, red sandstones with pebbly bands, dipping south-southeast at from 20° to 25°, occupy the coast and are cut by a number of basalt dykes. Note the different orientations of the latter.

8a. Here, where the path from North Corrygills leads down to the shore at Dunan, the raised beach feature widens out and there is a break in the long line of old sea-cliffs. Note the large granite erratic on the low raised beach. Portions of two higher terraces can be made out, the upper corresponding to the High Lateglacial Raised Beach (see pp. 63–65).

9a. Red sandstones with bands of breccia continue to occupy the shore. Note how at a point 125 m southeast of the mouth of the Corrygills Burn a 1·2 m dyke running NNE has cut across and shifted a 30 cm dyke directed ENE. Splendid examples of dune-bedding can be seen along this part of the coast.

10a. Note here the beginning of the "Mile Dyke" which runs along the foreshore in a southeasterly direction for about 1·3 km. Along the greater part of its course it forms a trench-like feature and can be seen in places to be steeply inclined to the northeast. It is an olivine-dolerite.

11a. About 450 m southeast of the mouth of the Corrygills Burn a felsite sheet, associated with pitchstone on its lower (northern) margin, crosses the shore in a west-northwesterly direction. Where best exposed on the shore it is some 6 to 7·5 m thick but it decreases rapidly in thickness seawards to a metre or so. It has been intruded along the bedding of the red sandstones, here dipping to the SSW at between 25° and 35°. Its upper margin is highly irregular and appears to be much more steeply inclined than the sediments. Its sill-like form is seen when traced inland, where it cuts the raised beach cliff and appears in both branches of the Corrygills Burn about 1·6 km to the west-northwest (Excursion 3, locs. 2 and 3). It is a tough dull greenish-grey rock crowded with spherical bodies up to 3 mm in diameter, which show a dark centre and often a dark border. A polished face shows the spherulites standing out as light-coloured spots against the dull groundmass of the rock. Each spherulite has a radial structure, due to the individual crystal fibres being set round a common centre. It should be noted that there is a good deal of variation in the extent to which the spherulitic structure is developed; sometimes the spherulites are closely packed and sometimes there is a considerable development of irregular granular aggregates which give the rock in such cases a spotted appearance. The lower part of the sill is banded and fissile.

This spherultic felsite is a well-known rock in geological literature and has often been described (see for example, McCulloch 1819, vol. ii, pp. 404–406; Teal 1888, plate 39, fig. 1; Tyrrell 1928, pp. 212–230).

The pitchstone which accompanies the felsite on its northern or lower margin has been described as the "small pitchstone" of the

Corrygills shore as opposed to the "great pitchstone" of locality 15a. Where the sill traverses the raised beach cliff the pitchstone is about 60 to 90 cm thick; it appears again in two pockets in the upper surface of the sandstone underlying the sill, near high-water mark. Tyrrell (1928, p. 212) describes the pitchstone in one of these pockets as having "a wrinkled appearance like ropy lava". The pitchstone is also spherulitic and presents a strikingly beautiful appearance under the microscope, the spherulites and indeed their individual fibres standing out against the dark glassy matrix which forms the bulk of the rock.

12a. Quartz-breccias and coarse sandstones containing numerous but generally small agate pebbles occupy the coast between localities 11a and 12a (overenthusiastic collecting over the years has resulted in the removal of a great deal of the agate; visitors are requested not to collect here, merely to note the presence of the agate pebbles, if they can find any, and pass on). These rocks may be regarded as forming the highest member of the Brodick Beds. In addition to the agates they contain pebbles of acid igneous rocks (granite, felsite, etc.) of Old Red Sandstone types, as well as the usual assortment of fragments of quartzite, vein-quartz and schist. Barrett (1925, p. 268) has recorded the occurrence of pebbles of andesitic lava with the agates. The assemblage of pebbles suggests derivation from a land area composed mainly of Dalradian schists but also in part of Old Red Sandstone and possibly Carboniferous rocks. Agates are common infillings of the vesicles (cavities in lavas produced by escape of gases) in the lava flows of Lower Old Red Sandstone age in Ayrshire and elsewhere on the mainland.

13a. Examine sections in red sandstones dipping southwest or south-southwest at an angle of about 15°. Note the relatively small size of the included pebbles as compared with those found farther north along the coast. The beds from here onwards to Lamlash are regarded as falling into the Lamlash Beds (see table p. 37). A pale-coloured dyke of quartz-porphyry traverses the raised beach cliff but appears to end against the "Mile Dyke". It runs in a general easterly direction and may be connected with the intrusion of quartz-porphyry forming Dun Dubh (Excursion 3, loc. 5).

Fine views may be obtained from this part of the shore of the northern escarpment of the great Clauchland Hills sheet.

14a. Note here the yellowish-weathering dyke of quartz-porphyry 1·5 to 5·5 m broad, running in an east-northeasterly direction.

15a. This is the locality of the well-known pitchstone sheet of the Clauchlands shore. It is sometimes referred to as the Corrygills pitchstone or as the "great pitchstone" on the Corrygills shore in contra-distinction to the thinner pitchstone intrusion accompanying the spherulitic felsite of locality 11a. It has frequently been mentioned or described (eg. Tyrrell 1928, pp. 213, 230, and references therein). The pitchstone is exposed at the base of the crags, where it is inclined to the SSW at much the same angle as the sandstones associated with it (30°). Note should be taken of its slightly transgressive character, of its rudely columnar structure, and of the somewhat decomposed bands present near its margins, especially the lower one. Its thickness is generally given as 3·7 to 4·3 m but it may reach as much as 6 m. It can be followed along the foot of the cliff for some 135 m and reappears on the foreshore, 185 m or so farther south, for a distance of 15 m, approaching within a metre or so of the "Mile Dyke". It is a dark, bottle-green, glassy rock, with a characteristic resinous lustre. Under the microscope it shows a pale yellowish, glassy base through which are scattered needles or microlites and feathery or arborescent growths composed of very small crystallites. A few well-shaped crystals of quartz are also present. There are thus four aspects to this rock; the crystals, the needles or microlites, the much smaller crystallites, and the glassy base.

Pitchstone flakes have turned up at a number of prehistoric sites in Scotland as far afield as the Tweed Valley and Caithness. Many of these closely resemble Arran pitchstone, in particular the "great pitchstone". Rock from this locality, and probably from some of the other pitchstone localities in Arran, was evidently traded extensively by man as far back as the third millenium B.C.

16a. The rocks here are all red sandstones belonging to the Lamlash Beds. As they approach the margin of the Clauchlands sheet their dip increases to 50°, they lose their red colour and at the contact they are indurated. Note the large granite erratic on the shore.

17a–18a. Between these localities the Clauchlands sheet occupies the shore continuously. At 17a the contact is seen to be partly vertical and partly to dip to the SSW. Against the country rock, as well as around included blocks, basaltic and doleritic "contact rocks" (see p. 46) are developed. Traced upwards, these give way to much coarser types: some 6 m above the contact the sheet is formed of a gabbro carrying very coarse grained patches with large crystals of black augite and cut by pale-coloured veins. At locality 18a the upper margin of the intrusion is concordant with the overlying baked sandstones which dip at 45° to the south-southwest.

The Clauchlands sheet may be described as a coarse-grained analcime-olivine-dolerite (crinanite), carrying, in addition to the two minerals named, labradorite feldspar and augite, together with apatite and iron ores. There is much variation in the proportion of mafic minerals (olivine and augite) and the lighter constituents (feldspar and analcime), a variation correlated with changes in the colour of the rock. In some exposures dark-green olivine is extremely abundant. The analcime is often replaced in part or completely by zeolitic minerals due to secondary changes which took place after the main consolidation of the rock.

19a. The sections along the shore at or near this locality are red, generally flaggy and cross-bedded sandstones with pebbly bands. Note how the sandstones become bleached along their contacts with the northwesterly basalt dykes.

Excursion 3: **Corrygills and the Clauchland Hills**

(Figure 7, localities 1 to 12)

The purpose of this excursion is to continue the examination of the Tertiary dykes and sills of the Corrygills district and in

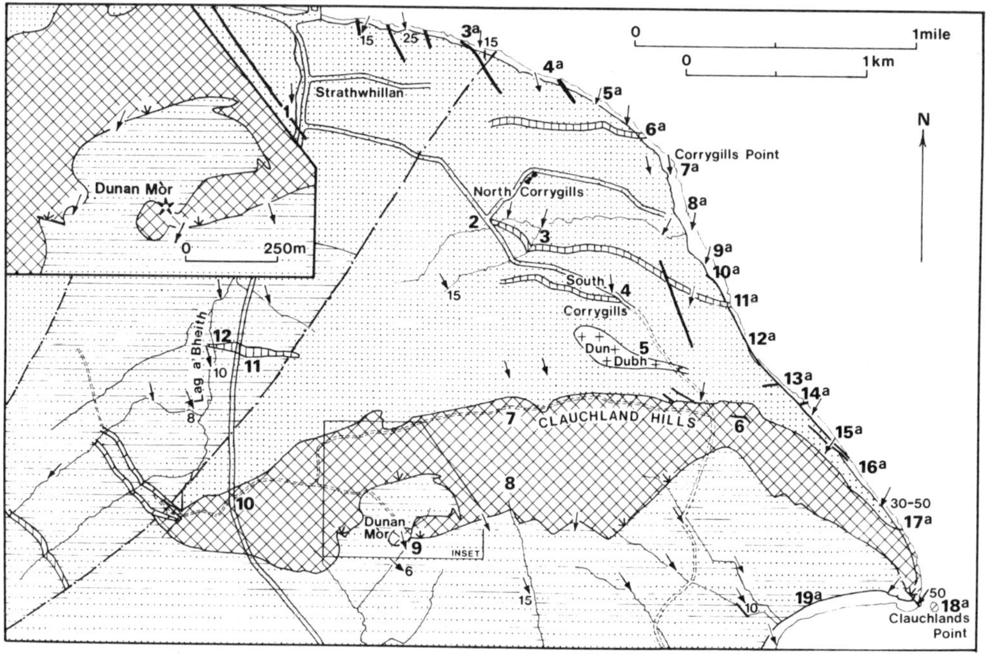

FIG. 7. Geological sketch-map of the Brodick–Lamlash area to illustrate Excursions 2 and 3. For key to map see Fig. 5, p. 78.

particular to study in more detail the textural variations and mode of intrusion of the massive sheet of olivine-dolerite forming the Clauchland Hills. The coastal sections, between Brodick Pier and Clauchland Point, have already been described in Excursion 2.

1. The red false-bedded pebbly sandstone exposed in the lower part of the Lag a' Bheith (sometimes referred to as the Strathwhillan Burn) have an inclination slightly west of south of between 15° and 20°. Dark-red sandstones, seen almost opposite the point where the Strathwhillan road leaves the main road, contain layers with ashy material and may be referred to the central part of the Brodick Beds (see Excursion 2, loc. 3a).

Almost due west of the junction of the Lamlash and Corrygills roads an olivine-dolerite dyke, 15 m or so thick, crosses the burn in a northwesterly direction and can be traced at intervals until it cuts the Brodick foreshore at Invercloy (fig. 6). The baked and indurated sandstone at the southwest contact stands up to form a small waterfall which tends to be obscured by a heavy growth of *Rhododendron*.

Follow the Corrygills road for a distance of about 1 km and note that the drift is generally thin there being outcrops of red sandstone and siltstone in the ditch on the south side of the road. Two dolerite dykes also outcrop in the ditch as they cross the line of the road.

2. Here the road forks, one branch leading to North Corrygills and the shore at Dunan, the other to South Corrygills. Pitchstone associated with felsite outcrops in the northern branch of the Corrygills Burn; however, a better section of this intrusion occurs in the southern branch of the burn at locality 3.

3. About 90 m downstream from the road a felsite-pitchstone sill forms a series of small waterfalls in the burn. Careful examination of the outcrops reveals massive, spherulitic felsite underlying indurated sandstone and underlain by 60 cm of dark-green pitchstone. This locality should be compared with the section of the same sill encountered on the Corrygills shore (Excursion 2, loc. 11a). Just north of the pitchstone a basalt dyke crosses the burn.

4. Here, thin-bedded red sandstones are exposed and traces of a spherulitic felsite are seen by the roadside: locally associated with pitchstone, it appears to follow a parallel course to that of locality 3. It is a very striking rock both in hand specimen and under the microscope.

5. The prominent hill known as Dun Dubh, about 215 m in height, is formed of a massive intrusion of quartz-porphyry some 137 m across at the widest part. Note the columnar structure of its northern face. Eastwards it thins out to a narrow tongue. The rock itself carries crystals of quartz and feldspar set in a yellowish or yellowish-grey, minutely crystalline groundmass showing flow-structure. Its field relationships suggest that it is a more or less vertical, plug-like intrusion. The dyke of quartz-porphyry cutting the cliffs and the foreshore to the east (Excursion 2, loc. 13a) may be connected with the Dun Dubh intrusion.

6. The hill-top known as Dun Fionn lies a little east of where the track from South Corrygills crosses the Clauchland Hills. Rising abruptly from the shore to a height of about 163 m it is a magnificent viewpoint commanding a wide panorama. Hardly a trace of the fort that at one time existed is now left. James Bryce, writing in 1872 says (p. 79) "A low mound, enclosing an elliptic space 40 yards by 16, is seen round the summit, but nothing whatever is known of the history of the fort". Today it is difficult to detect even so much.

On the southern slopes of Dun Fionn a spherulitic pitchstone sill cuts the dolerite of the Clauchland Hills, possibly a continuation of the lower one described below (Gunn 1903, p. 93).

The path which follows the crest of the Clauchland Hills westwards should be followed for about 300 m. A short deviation northwards from this point should be made to look for and examine the lower contact rocks of the Clauchland intrusion and, farther downhill, two intrusions of pitchstone (of Corrygills type). Both are sill-like and form isolated ledges a little below the scarp features of the great Clauchland Hills dolerite. The lower one is some 4·5 m in thickness, the upper one about 3·5 m, but neither can be traced far laterally. Both are dark-green rocks showing under the microscope a pale yellowish-brown glassy base crowded with numerous greenish needles and feathery growths. The greenish needles or microlites are probably hornblende (Tyrrell 1928, p. 230).

7. Continue along the crest of the hills as far as the summit cairn noting the coarse dolerite exposed in the rocky crags. The contact rocks can be looked for, but may be difficult to find, in the low feature often developed a short distance north of the line of crags.

8. Cross the outcrop of the intrusion to this locality *via* a path on the west side of the dense plantation. The rock here is a coarse-grained analcime-olivine-dolerite with irregular pockets and veins of very coarse-grained crystalline material. Farther south, and close to the upper margin, which side-steps abruptly along the line of a stream, alternations of fine and coarse-grained dolerite occur, intersected in places by narrow tachylitic veins. To the east of the stream an ENE-WSW striking rock face, with adherent patches of basalt, appears to plunge at 40° or so to the south-southeast.

9. Here, a short distance southeast of the Dunan Mòr cairn, dolerite associated with dense basalt plunges steeply to the south-southeast. In the adjacent streams the intrusive rock abuts on sandstones. Farther to the northeast a leaf of the intrusion overlies the sediments. An attempt should be made to examine fully the rocks in and surrounding the large embayment of sediments. "Contact rocks" dipping south-southeastwards through highly baked sandstone (locally with paramorphs after tridymite) occur along the northern margin of the embayment.

Those interested in archaeological remains may wish to examine the two chambered burial cairns (or rather the remains of them) of Dunan Mòr and Dunan Beag. Both were excavated in 1909 and the following particulars are taken from the detailed account of the sepulchral remains given by Bryce (1910, pp. 75-82):

Dunan Beag (fig. 8). This cairn, standing about 400 feet (c. 120m) above sea level, measures 121 feet (37m) in length from north to south and is 65 feet (20m) in breadth. At its northern end

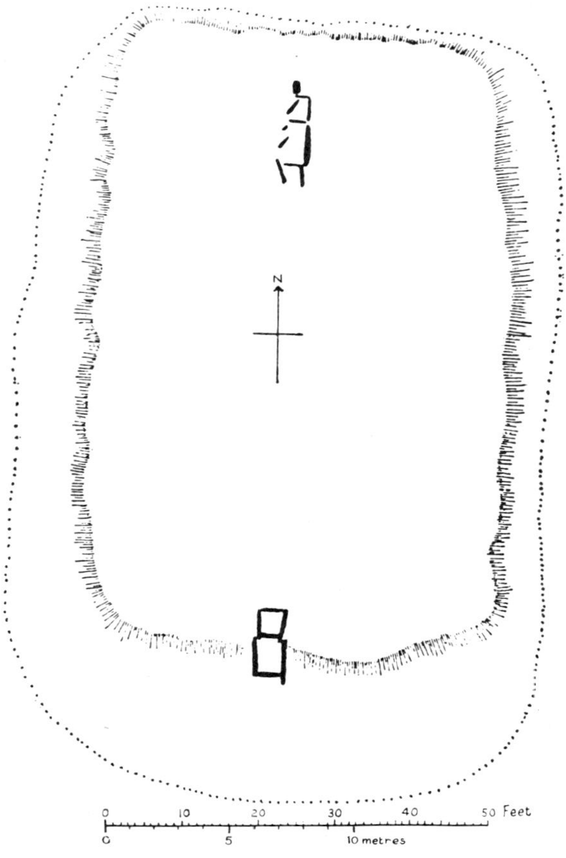

FIG. 8. Plan of Dunan Beag Cairn, Lamlash; from *The Book of Arran*, 1910, p. 76.

there was a chamber, much dilapidated, with three compartments 3·5 to 4 feet (over a metre) deep; these were found when opened out to be filled with stones and earth. They had certainly been rifled at one time and the only trace of interment was a small piece of burnt bone. Relics recovered comprised a fragment of unornamented pottery composed of dark ware, a flint flake showing some signs of working, and some flakes of Corrygills pitchstone.

At the south end of the cairn was a second chamber of two compartments, possibly all that was left of a larger structure. The first compartment measured 5 feet (1·5m) in length by 3ft. 3in. (1m) in breadth; the second measured 3 feet (0·9m) by 2ft. 10in. (0·86m). Each was filled with soil and stones and the floors were covered by a layer of black earth, containing many fragments of charcoal. The first compartment contained the unburnt remains of two adult persons, lying in a doubled-up position. The second compartment yielded another unburnt interment. A number of animal bones, chiefly of the ox, were also found, while the relics recovered comprised a flint flake, some fragments of Corrygills pitchstone, a small piece of dark-coloured pottery, portions of an ornamented urn of red ware, and parts of a jet necklace. The urn fragments represent a small vessel of the beaker type (see p. 71).

Dunan Mòr. The cairn, which lies about 100 feet (30m) higher up the slope, is a circular structure with a diameter of 78 feet (24m). When excavated in 1909 the remains of three chambers arranged radially round the margins were disclosed but all in a very ruined condition. Practically nothing remained of the chamber on the northeast margin. That on the west side was 22 feet (6·7m) long, with three compartments each 3ft. 2in. (about 1 m) deep. No trace of interments was found and no relics, suggesting that it must have been completely rifled at one time. The chamber at the south end was the best preserved of the three and showed two compartments, here again probably portions of a larger structure. The floor of each compartment was covered with a layer of black earth containing some pieces of charcoal and fragments of burnt bone.

The relics recovered comprised a large, coarsely made vessel of red pottery, a flint knife or scraper, three flint flakes, and some fragments of Corrygills pitchstone. The flint knife measured 8 inches (20 cm) in length by 3·2 inches (8 cm) in breadth.

10. Fine-grained dolerite is exposed on the east side of the Brodick–Lamlash road at this locality. It is perhaps best seen a little south of the circle of standing stones. The dolerite here represents the thinned out western extension of the Clauchland Hills sheet.

The standing stones referred to stand on a slight elevation on the east side of the road and consist of four massive granite blocks delimiting a circular arc with a diameter of 5·2 m. The blocks are round-topped and vary in height from 0·56 to 1·17 m. In 1961, James Bryce (1872, p. 216) "found within the circuit of the four blocks, and at a small depth, a cist 26 inches (0·66 m) long, 11 inches (0·28 m) wide and 10·5 inches (0·27 m) deep, cut out of the solid sandstone rock, and covered by a lid. In it there were bone fragments and black earth, in the soil over it some rude flint arrow-heads". Later, however, Bryce (1910, p. 122) described the remains found in the rock above the cist as "a few flint flakes and a flint implement".

A magnificent view of the north Arran hills can be obtained from this point (Plate 1).

11. At this locality a small quarry has been opened in a generally fine-grained, pale-coloured, slightly yellowish felsite about 7·6 m thick. Note the platy jointing of the rock. Spherulitic structure is not obvious in hand-specimen except along a narrow marginal zone. The intrusion shows on both edges a thin band of pitchstone. This felsite has been described by Allport (1872a, p. 542) and more recently by Tomkeieff (1946 and 1961).

12. A dark-green pitchstone is exposed here in the old Brodick–Lamlash road and in the Lag a'Bheith burn close at hand. It is probably a continuation of the quarried rock described above.

Excursion 4: **Corrie Shore**

(Figure 9, localities 1 to 12)

The main purpose of this excursion is to examine the Carboniferous rocks of Arran in an area that is readily accessible and where the succession of rock types is clearly displayed. The succession on the Corrie shore is illustrated diagrammatically in figure 10. The thickness of the Carboniferous sedimentary rocks present is very approximately 295 m, made up of Coal Measures, about 75 m; "Millstone Grit" (Passage Group), perhaps 9 m; Carboniferous Limestone Series, about 120 m; and Calciferous Sandstone Series, about 90 m. A group of basaltic lavas, approximately 115 m thick, is intercalated near the base of the Calciferous Sandstone Series.

The excursion also offers an opportunity of examining the junctions of the Carboniferous strata with the underlying Old Red Sandstone and the overlying Permian.

1. Examine at this locality the exposures of red, cross-stratified and often coarse-grained sandstones belonging to the Upper Old Red Sandstone, dipping to the southeast at angles of between 24° and 30°. Note that the dips steepen locally in the vicinity of small crush-lines or faults. At the roadside there is a huge erratic block of granite known as the Cat Stone (Gaelic, Clach a'Chait); its dimensions are approximately 4·3 m high, 5·5 m long and 2·75 m wide.

2. A band of reddish conglomerate, 21 m or so thick, outcrops here, and is followed in upward succession by a series of red cross-stratified sandstones, sometimes showing on their upper surfaces many irregular ridges which may represent infilled mud-cracks. These sandstones continue along the shore until another red conglomerate band is reached underlying the lowest cornstone bed (*b*) of locality 3. The Upper Old Red Sandstone in the vicinity of localities 1 and 2 shows a derivation from the south to south-southeast.

3. Examine the upward succession of rock types exposed at this locality and note their variable and often disturbed character.

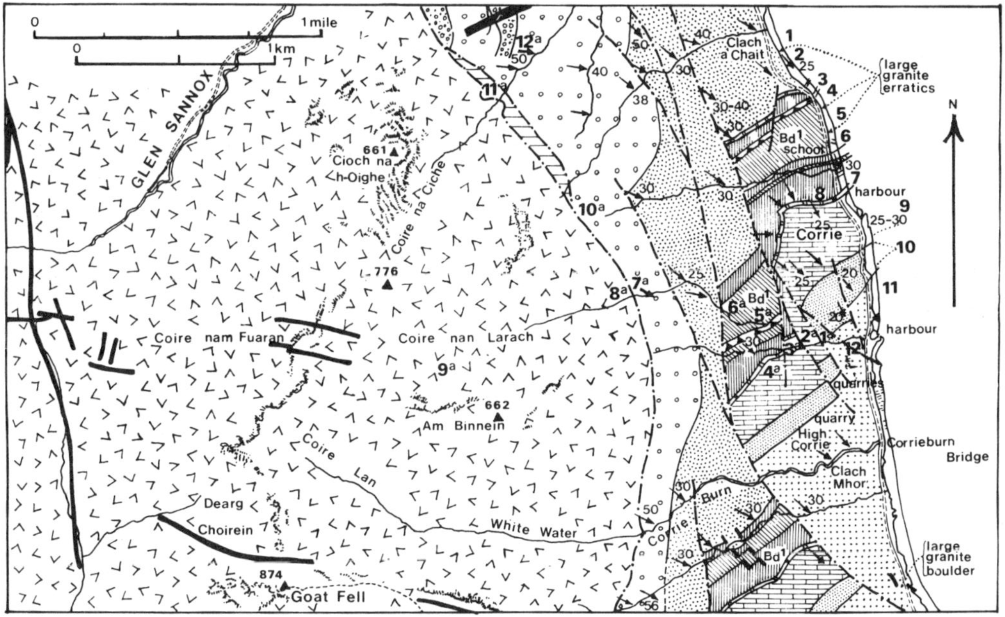

FIG. 9. Geological sketch-map of the Corrie area to illustrate Excursions 4 and 5.
For key to map see Fig. 5, p. 78.

Succeeding the red sandstone at locality 2 is a second band of red conglomerate 12 to 15 m thick, which is followed by some 9 m of strata which include three bands of nodular cornstone. The sequence from above downwards is approximately as follows:

(*e*) Cornstone, pale-coloured, sandy, nodular, about 75 cm.
Gap in succession, say 90 cm.

(*d*) Cornstone, nodules in red shaly matrix, about 4·5 m.

(*c*) Mudstones, red, with bands of pale-grey sandstone, about 1·2 m.

——Line of disturbance——

(*b*) Cornstone, pebbly, associated with calcareous sandstone, 1 m.

(*a*) Conglomerate, massive, reddish, up to 15 m.

Gunn (1903, p. 38) regarded bed (*a*) as the highest member of the Upper Old Red Sandstone and the junction with the pebbly cornstone (bed *b*) was taken by him as "clearly an unconformable one". The writer considers that the junction between the two systems may be drawn more fittingly just above (*e*).

4. The highest of the three cornstones is succeeded by a series of generally pale-coloured sandstones inclined to the southeast at angles of about 25° to 28°. These sandstones, 6 m or so thick, are followed by a coarse agglomerate of at least 15 m in thickness made up of fragments of basalt lava similar in type to the lavas of localities 5 and 6.

5. Examine at this locality the lower part of the lavas (Volcanic Group of Calciferous Sandstone Series), showing dark-blue to purplish olivine-basalts with phenocrysts of augite, olivine and occasional plagioclase feldspar. Good examples illustrating weathering into spherical masses can be seen here. Note the large granite boulder known as Clach an Fionn on the shore side of the Corrie road.

6. The upper part of the lava succession is made up of purplish olivine-basalts, generally fresh and providing excellent specimens for microscopic examination. A section of a sample collected from the shore opposite the schoolhouse is reproduced in plate II (fig. 9). A search should be made for the slaggy, decomposed,

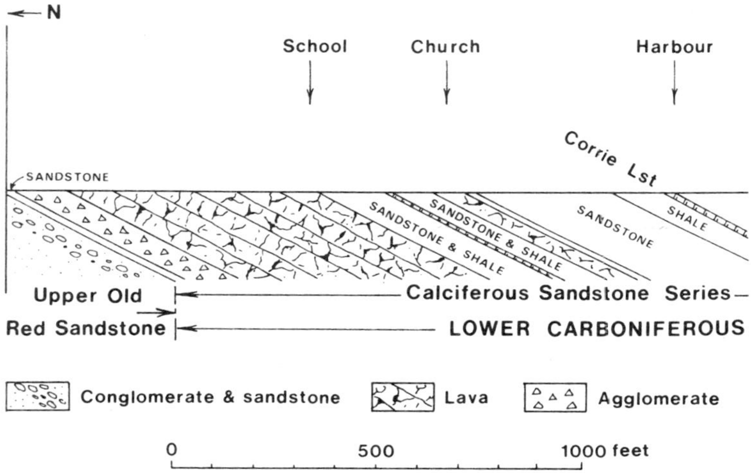

FIG. 10. Diagrammatic cross-section

amygdaloidal tops of individual flows. Thin ramifying veins of calcite are common near the top of the group.

The lavas are overlain by reddish sandstones and shales, including a 46cm red limestone bed which dips southeast at about 25°. Some of the sandstones carry small fragments of volcanic ash. These beds are followed in turn by a band of purplish basalt lava 3·0 to 3·6m thick, much veined by carbonates.

7. There is a gap in the shore-section above the last lava of locality 6, a gap possibly representing some 9 to 10·5m of soft beds. The next beds exposed are massive sandstones, pale-grey in colour but locally with a reddish staining. At one time they were quarried along their outcrop both on the shore and for some little distance inland. Overlying the sandstones come soft reddish mudstones, which, however, are poorly exposed and must be looked for carefully. One sample obtained by the writer contained

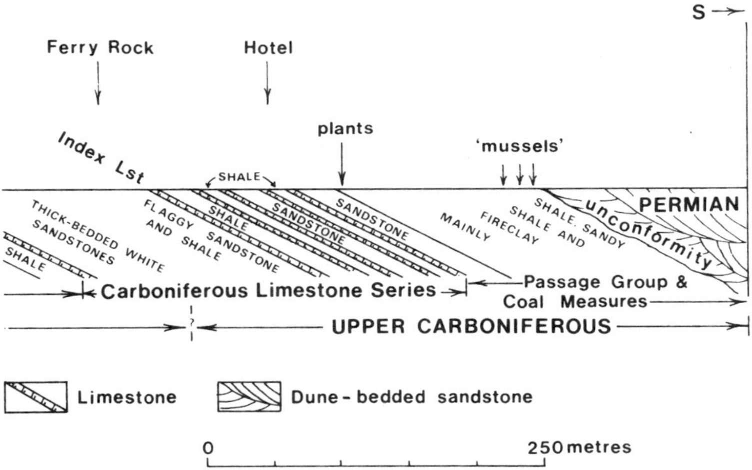

rock succession on the Corrie Shore.

a fragment of a ribbed brachiopod shell. The shales underlie the thick Corrie Limestone of locality 8.

8. The Corrie Limestone is the basal member of the Carboniferous Limestone Series in Arran (see table, p. 31). It has been excavated to form the old harbour at Corrie, and has also been extensively quarried and mined along its outcrop inland for a distance of nearly 365 m. The present face shows the limits to which quarrying has been carried, and along it a number of day-holes ("ingaun e'es", as they used to be called) indicate entrances to the old underground workings. Good sections of the limestones and of the overlying beds can be seen at various points along the present face, and at the entrances to the mines, **but the visitor should on no account attempt to penetrate the galleries as there is a danger from possible roof-falls.** The limestone is said to have been 6 m thick, but only some 3·6 to 4·3 m of good limestone,

forming the upper part of the seam, is actually visible. It is overlain by a cover of reddish siltstones and shales, reaching in places a thickness of 9 m.

The limestone is divided into separate beds, or posts, by partings of reddish fossiliferous calcareous shales, and on the upper surfaces of these beds there are numerous shells of the large brachiopod *Gigantoproductus giganteus.* A layer packed with specimens of this species forms a striking feature of the roof of the old workings. Both the calcareous partings and the intermediately overlying shales contain a rich fauna including brachiopods, lamellibranchs, gastropods and cephalopods. A list of species obtained from the shales above the limestones is given in the Arran Memoir of 1928 (p. 62). The limestone was formerly worked and calcined for local use, as well as for export, and the remains of old kilns still exist close to the main road. A bulk sample representing the whole face exposed was analysed some years ago in the laboratories of the Macaulay Institute for Soil Research, Aberdeen, and showed 88·21% calcium carbonate, 1·3% magnesium carbonate and 10·29% insoluble residues (including 7·33% silica).

Note the two thin dykes cutting the limestone, one a little above the old lime-kilns, the second 45 m or so higher up.

9. Examine the strata that intervene between the Corrie Limestone and the lowest limestone of locality 10, that is, the strata which are regarded as corresponding to the Lower Limestone and the Limestone Coal Groups of the mainland succession. Their thickness may be estimated as about 76 m. Following upon the shales and limestones seen above the Corrie Limestone comes a series of thick white or faintly pinkish sandstones alternating with bands of shale. Some of the white sandstones are fine-grained and exceptionally pure, with a high silica content. They can be seen in the rough ground in the wood above Corrie Village, where there are overgrown quarries in which they were worked in the early part of the 19th century. Hard, compact sandstone is also exposed on the foreshore where it forms a large projecting mass of rock which was also quarried to supply large blocks for harbour

construction in the Isle of Man. The site of these old excavations was later used as a convenient access and landing stage for the ferry boats from the Brodick steamers and the part of the rock left and used as a jetty was then known as the Ferry Rock. The ferry ceased operations in 1939. The beds exposed above the sandstone on the shore include a few seat-earths, interbedded with sandstones, shaly sandstones and shales which extend southwards to a little beyond the Ferry Rock, but no trace of coal can be seen.

10. Note here the reddish limestone, 60 cm or so in thickness, which has been tentatively correlated with the Index Limestone at the base of the Upper Limestone Group of the mainland. It has yielded among other brachiopods, the form *Productus latissimus* together with small gastropods and a few lamellibranchs. Several higher marine limestones, ranging in thickness from 30 to 90 cm and separated by shales and occasional sandstone beds, are present. The strata including them extends southwards to the little bay in front of the hotel. All these beds are placed tentatively in the Upper Limestone Group. Owing to the lack of continuous exposures and to the occurence of small faults or crush-lines, their thickness can only be approximately estimated at about 42 m.

11. Under this locality number are included all the strata seen in the bay a little south of the Corrie Hotel. Concerning these Gunn (1903, p. 41) said: "The beds in this bay are classed with the Coal Measures, as they contain organic remains of Upper Carboniferous facies, but they and the equivalent beds elsewhere may possibly be Millstone Grit." As a result of the research carried out by Leitch during his revision of the Upper Carboniferous rocks of Arran (1942, pp. 141–154), they must now be placed in the Productive Coal Measures. The detailed section recorded by Leitch gives a thickness of 84 m for these Upper Carboniferous strata at Corrie, the lowest bed being a 9 m of massive white sandstone with a conglomeratic base. The sandstone may be equivalent to the upper part of the Passage Group, the conglomeratic base representing the unconformity of early Passage Group times which is known over much of Lanarkshire. The detailed section is reproduced below:

	m
Highly contorted sandstone, forming local top of Coal Measures	2·4
(Thin sill intruded here)	
Cross-bedded and slightly contorted sandstone	1·5
Red shales with "mussels"	3·0
Ribs of cross-bedded sandstone	1·8
Purple shales with "mussels"	3·0
Gap	?
Sandy shales	0·9
Red papery shales with "mussels" at base	7·3
Marl and shaly sandstones	3·0
Red and purple shales	4·0
Fireclay, irregular	2·1
Shales	2·4
(local unconformity)	
Purple shaly sandstone and marls with several fireclays	15·2
(local unconformity)	
Fireclay	4·9
Clayey, hummocky sandstone	3·0
Sandy shales	1·8
Marls or fireclay	2·7
Sandstone	3·7
Fireclay	1·8
Cross-bedded sandstone	6·1
Shales with plant-remains	0·8–0·6
White sandstone	0·6
Fireclay	1·8
Shales	0·6
Massive, white sandstone with conglomerate at base	9·1

This table shows that at three horizons in the upper part of the succession, beds of reddish mudstone yielded to Leitch non-marine lamellibranchs referable to the *modiolaris* zone of the mainland. The approximate distance of these horizons below the

Carboniferous-Permian junction and the "mussels" recorded from each are:

Red mudstones from 4 to 7 m below the junction:
abundant small oval shells referable to *Carbonicola aquilina* and *C. elliptica*.

Purple mudstones from 8·8 to 11·9 m below the junction:
Carbonicola aff. *aquilina, C.* cf. *phrygiana C.* cf. *regularis* and *Naiadites triangularis*.

Red papery mudstones a little lower down:
Carbonicola os-lancis and *C.* cf. *rhomboidalis*.

The range and distribution of the non-marine lamellibranch zones in Arran are briefly discussed on p. 35. It should be noted that while the *modiolaris* zone of the mainland succession includes a number of workable coal seams, there are no traces of coal anywhere in the Corrie section. Apart from the fossiliferous mudstones already mentioned, the sediments consist of thin-bedded sandstones, sandy shales, and fireclays or marly fireclays carrying rootlets. The sedimentary structures present, such as small-scale cross-bedding, contorted bedding, flow bedding and minor unconformites or non-sequences point to deposition on a deltaic platform. Leitch (*ibid.,* p. 151) has suggested that the absence of coals in the "Coal Measures" of Arran is "presumably due to the marginal position of these areas in relation to the main coal basin of Carboniferous sedimentation". It is certainly remarkable that coals of workable thickness (up to 1·7 m) occur in the Productive Coal Measures at Machrihanish, 43 km or so southwest of Corrie as well as on the Ayrshire coast, some 23 km to the east. It would seem that conditions at Corrie were never sufficiently stable to allow the growth of swamp-forests.

12. The unconformable junction of the Carboniferous beds with the Permian (Corrie Sandstone) at this locality should be carefully examined. Note the sudden contrast in rock types across the line of junction and that the line of junction is roughly parallel with the bedding of the Carboniferous sediments. Bailey (1926a) has described it as marked by scattered, minute pebbles of vein-quartz

and iron pan mixed with rounded quartz grains. It should be noted also that at Corrie the Permian rests on beds assigned to the *modiolaris* zone, while in the Merkland Burn at Brodick it overlies strata assigned to the higher *similis-pulchra* zone (see p. 35).

The opportunity should be taken to visit the old quarries in the Corrie Sandstone south of the village where the cross-bedding and the rounded quartz grains characteristic of aeolian deposits can be examined. Plate II, figure 1, is a drawing of a micro-section made from a specimen from the quarries here. These quarries supplied much building stone both for local use and export.

Excursion 5: **Locherim Burn and Coire nan Larach**

(Figure 9, localities 1a to 12a)

The main objects of this excursion are: (1) to continue the study of the Carboniferous rocks of the Corrie district: (2) to examine the junction of the Lower Old Red Sandstone and the granite in the Locherim Burn and the contact of the latter with the Dalradian schists in the Allt a'Chapuill: (3) to see the lateral moraines and other glacial phenomena in Coire nan Larach.

The Locherim Burn rises in Coire nan Larach and enters the sea at the old quarry to the south of Corrie village. About 350m up from its mouth the stream forks to form the North and South Locherim Burns, the former being the main branch. The Allt a'Chapuill rises on the western slopes of Cioch na h-Oighe and joins the Glen Sannox Water near the old barytes mines.

It may be pointed out that the route to be followed on this excursion traverses the 300m (1,000 foot) Platform and the opportunity should be taken to note how this feature slopes gradually upwards from its outer margin at about the 245m level to where the steep ground begins at about 360m.

1a. Examine the section of Coal Measures exposed in the lower part of the Locherim Burn about 180m up from its mouth. The section is a short one as the strata belonging to the Coal Measures are faulted in the east against the Corrie Sandstone and on the west against rocks which fall into the Upper Limestone Group. The section shows in descending sequence, sandy marls, fine red shales, thin-bedded shaly sandstone and a fireclay with rootlets. The red shales have yielded plant-remains (leaflets and pinnules of *Mariopteris* and *Neuropteris* and some non-marine lamellibranchs (mussels). Leitch (1942, p. 144) refers the mussels to *Carbonicola os-lancis* and *Anthracomaya modiolaris* and correlates this section with the top part of the Corrie shore section (see Excursion 4, loc. 11). He regards the mussel band here as the equivalent of the lowest of the three mussel-bands of the Corrie shore.

2a. Examine at this locality the section in strata belonging to the Upper Limestone Group exposed on the south side of the burn, almost opposite the point where it forks. The strata seen here are, in downward succession, thin-bedded siltstones alternating with bands of shale, shales containing plant remains, fossiliferous calcareous shales and a thin limestone. The plant-bed has yielded a number of species among which *Lepidodendron veltheimianum* is the most abundant, while the fauna of the underlying calcareous beds includes brachiopods, lamellibranchs and gastropods. Here again the section is a short one, the strata being separated by faulting from the Coal Measures of locality 1a to the east and from beds associated with the Corrie Limestone to the west.

3a–4a. At these localities some of the beds associated with the Corrie Limestone can be examined. The position of the Corrie Limestone itself can only be inferred, although there are indications that quarrying operations have at one time been carried on. Note especially the occurrence of white, fine-grained very pure sandstone exposed in the south branch of the Locherim Burn a short distance above the assumed outcrop of limestone. A very similar sandstone occurs close by, above the Corrie Limestone at Corrie itself (Excursion 4, loc. 9).

5a. Return to the main branch of the Locherim Burn and at this locality examine carefully the rocks exposed in the gorge cut in the Calciferous Sandstone Volcanic Group. The Group here consists of lavas, together with a number of bands of agglomerate and tuff; a few thin bands of sedimentary material are also present. If this section is compared with the section of the same Group on the Corrie shore it may be noted that it contains a considerably greater proportion of rocks of fragmental origin, agglomerate and tuff, than the latter.

6a. The upstream or western margin of the lavas is defined by a fault seen to cross the burn at this locality in a southerly direction. Along part of the fault-line a thin dyke has been intruded, forming a waterfall. The fault brings the volcanic rocks against conglomerates belonging to the Upper Old Red Sandstone. Upstream from the fault good sections in red, often gritty, sandstones of the same formation are exposed, dipping to the east-southeast at about 25°.

The Upper Old Red Sandstone is shown (fig. 9) as separated by a fault from the Lower Old Red Sandstone, but it should be pointed out that between Corrie and Brodick the nature of the boundary between the two formations is very uncertain. There is certainly no good evidence for an unconformity here.

7a. Examine sections in purplish-red sandstones assigned to the Lower Old Red Sandstone. Note the steep dips and how the sandstones lose their red colour and become progressively harder as the granite margin is approached.

8a. Examine the exposures on both sides of the Lower Old Red Sandstone-granite boundary. The sandstones of the former have been baked and indurated and altered to a hard, shiny quartzite-like rock, tinted a faint green colour through the development of epidote. The granite near the junction is moderately coarse-grained.

It is well worth extending the excursion route from this point southwards, for about 1·2km, in order to examine another junction of the granite and the Lower Old Red Sandstone exposed in the Corrie Burn (White Water). Here the faulted junction is clearly seen and the Old Red Sandstone beds have been indurated, altered and epidotised for a distance of roughly 37m. Movement along the actual junction has given rise to a narrow zone of mylonite (see Bailey 1926a, p. 481).

9a. Coire nan Larach. Note the splendid example of a lateral moraine on the north side of the Am Binnein ridge. The surface of the moraine is strewn with large granite blocks. Note also the general covering of morainic drift on the corrie floor. The high rugged ground at the head of the corrie (810m) forms part of the granite ridge joining Cioch na h'Oighe to Goatfell.

10a. Examine small exposures of Lower Old Red Sandstone beds near the granite margin. These consist of shaly beds showing alteration to hardened, greenish-grey rocks in which traces of the original bedding are left. Just north of this locality there is a narrow strip of Dalradian grits bordering the granite for a distance of about 800m. The actual contact of the two rocks is seen in Allt a'Chapuill at locality 11a.

11a. This locality merits careful study. The actual junction of the granite and grits is sharp but the line of contact is somewhat sinuous. The grits are much indurated and have lost their schistose appearance, as recrystallisation, accompanied by the development of new minerals such as biotite has taken place. The granite close to the junction is fine-grained with phenocrysts of feldspar.

12a. Examine the stream section at this locality in chocolate-red and purplish sandy mudstones referred to the Lower Old Red Sandstone. These strata are separated from the altered Dalradian rocks of locality 11a by an important fault, not seen however, in the stream. Note the east-southeast dip of the beds at angles of almost 45°.

Below this section and about the 140m level there is an alluvial flat beyond which the stream turns sharply at right angles to its old course to flow northwestwards and join the Sannox Water.

The opportunity should be taken to examine the exposures of the broad west-southwest trending dyke which crosses the Allt a'Chapuill a short distance below the alluvial flat. This dyke can be traced for a distance of about 1·2km; part of its outcrop is shown on figure 9, part on figure 12. The rock is a coarse-grained quartz-dolerite described by Kynaston (in Gunn 1903, p. 173) as showing under the microscope "more or less lath-shaped turbid plagioclases, grains of pale augite of earlier consolidation than the feldspar, patches of pale greenish (sometimes yellowish in the more weathered part of the rock) secondary fibrous amphibole, with some chlorite and occasionally a little brown hornblende associated with the augite. There is also some interstitial alkali-feldspar and quartz". The intrusion is probably one of the late Carboniferous suite of quartz-dolerites.

Excursion 6: **Glen Sannox and North Glen Sannox**

(*Figure 11, localities 1 to 22*)

The main objects of this excursion are: (1) to traverse the Lower Old Red Sandstone succession seen in Glen Sannox and on the hillsides to the north, including a visit to the old barytes mines; (2) to study the exposures of Dalradian and Ordovician rocks on the south side of North Glen Sannox and in the North Sannox river itself; (3) to examine the sections of the coarse granite in Coire nan Ceum and the granite-schist contacts at selected localities; (4) to traverse the faulted outcrops of Lower and Upper Old Red Sandstone between North Glen Sannox and the Fallen Rocks.

The excursion begins where the path to Glen Sannox leaves the Corrie–Loch Ranza road. The path leads past an old burial ground, at one time the site of a small chapel dedicated to St. Michael. An ancient sculptured stone showing a rude effigy of the saint is built onto the outside of the graveyard wall to the left of the gate.

Before going on to locality 1, it is worthwhile lingering for a short time at some point giving a good view of the glen with its rugged rampart of encircling hills; the long southern face of Suidhe Fhearghas to the north; to the south Cioch na h'Oighe with the deep corrie known as the Devil's Punchbowl on its eastern side; and, at the head of the glen, Cir Mhòr.

1. At this locality sections in the Sannox Burn show fine-grained, purplish sandstones and hard, red, sandy mudstones, sometimes slightly calcareous, assigned to the Lower Old Red Sandstone. These are inclined nearly due east at angles of between 33° and 50°. Two dykes 2·4 m and 1·2 m thick, respectively, cut the sedimentary rocks; these are both badly decomposed basalts in

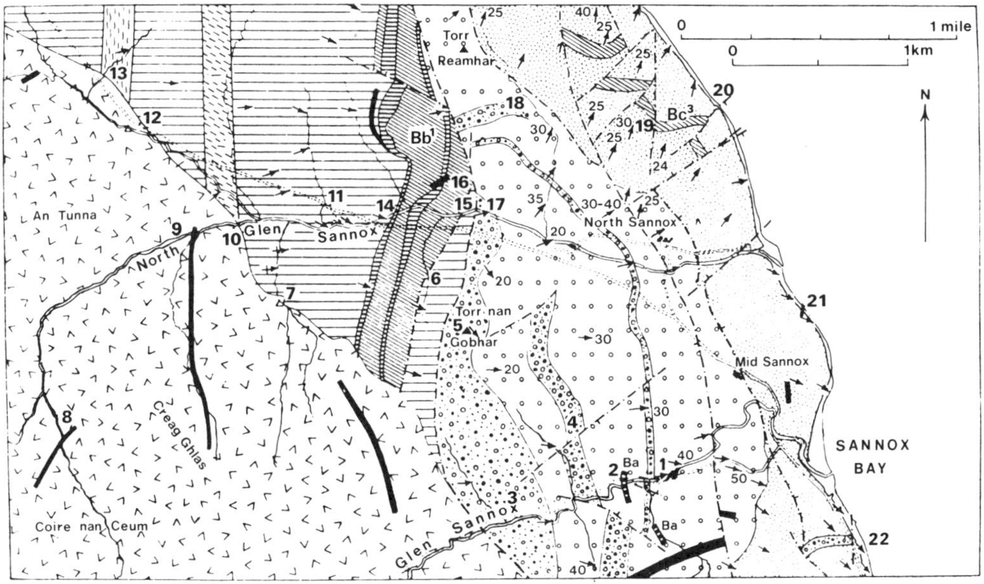

FIG. 11. Geological sketch-map of the Glen Sannox and North Glen Sannox area.
For key to map see Fig. 5, p. 78.

which the original porphyritic crystals of olivine and augite have undergone alteration and are now largely replaced by pseudomorphs. They belong possibly to the Lower Carboniferous period of vulcanicity. Note the band of quartzite conglomerate crossing the stream a little higher up.

2. *Barytes mines*. Barytes (barium sulphate) was worked here, both by surface excavation and shallow shafts, in the middle of the nineteenth century (*circa* 1836–1862), and again on a more extensive scale from 1919 to 1939. The main vein crosses the Sannox Water in a northerly direction and is about 1·5 m thick, steeply inclined to the west. During the later workings shafts were sunk on both sides of the burn and the ore followed downwards to a depth of about 9 m. Both downwards and laterally, however, the workings entered unproductive ground and they were abandoned in 1939. The barytes in the worked area occurred as a series of pockets or lenticles aligned along fault-fissures trending north or northwest and it is said that the richest deposits were found at the intersections of these. The crushing and screening plant was situated near the shafts and the ore was taken by a light railway with a self-acting conveyor to the loading pier at the mouth of the Sannox Water.

Exploratory mines were also driven northwards as far as the small veins exposed on the north hill (see loc. 4), but no important occurrences were found in the ground between the latter and the main field. Attempts were also made to reach the barytes vein seen crossing the little burn (Allt a'Chapuill) which joins the Sannox from the south, but again without success.

The output from the Sannox mines increased from 300 tons in 1920 to 8,693 tons in 1934. Barytes, known from its high density as "heavy spar" has many uses in industry, particularly as a component of drilling mud in oil wells, as well as being an important source of barium salts.

The country-rocks in the vicinity of the old shafts are fine-grained purplish sandstones, often clayey and slightly calcareous, and red mudstones. Abundant material is available for examination in the waste-heaps.

The broad dyke-like intrusion of dolerite south of the barytes mines is included in the itinerary for Excursion 5 (loc. 12a).

3. From the mines follow the rough track up the glen to locality 3. Examine here the exposures of massive conglomerate containing numerous rounded boulders and pebbles of quartzite as well as boulders of igneous rock. The latter are in the main hornblende-andesites and represent material derived by denudation in Lower Old Red Sandstone times from a contemporaneous volcanic series (see Excursion 1, loc. 16). A short distance beyond this locality the granite margin is reached, but the actual junction is not seen. It should be noted that the Dalradian schists are not present along this part of the granite margin (cf. loc. 5).

4. Cross the drift-covered hillside in a northeasterly direction to locality 4. Here there are fine exposures of a massive conglomerate made up mainly of quartzite pebbles and forming a broad belt some 120 m wide running in a north-northeast direction. Examine the types and shapes of the included pebbles and boulders and note also the thin barytes veins traversing the conglomerate. These veins, seldom exceeding 60 cm in width and very impersistent, were opened out and worked for a short time after the closing down of the main mines at locality 2.

5. Follow the conglomerates northwards from locality 4 for a distance of about 500 m and then turn westwards to Torr nan Gobhar (250 m). This little hill is built in the main of red and grey conglomerates overlain to the east by purplish, flaggy sandstones and hard mudstones. To the west the conglomerates are brought against the schistose grits of the Dalradian along a line which has been suggested to be a continuation of the Highland Boundary Fault (see p. 27).

The schistose grits form a prominent feature in the topography here, extending in a east-west direction for some 180 m and stretching from the North Sannox Water southwards for about 1·3 km until truncated by the granite. Graded bedding seen at a number of exposures indicates that the succession in this belt of schists is an upwards one from west to east.

6. At this locality the visitor passes westwards across the

southern part of the zone of black shales, cherts and lavas, of possible Arenig (Lower Ordovician) age, shown (fig. 11) as running in a north-south direction across North Glen Sannox. In a little burn at this point there is a section showing Dalradian schists separated from "Arenig" lavas by less than a metre of broken rock. Both schists and lavas are somewhat shattered, and Anderson and Pringle (1944, pp. 86–87 and fig. 1) have drawn a fault between them.

The westward route from locality 6 to locality 7 traverses the outcrop of the "Arenig", the rock groups met with being:

Upper lava
Thin band of black shales and cherty mudstones
Lower lava
Thin band of black shales.

The succession can, however, be best studied in the North Sannox Water (locs. 14–15) and may be passed over rapidly here. The Dalradian schists following the "Arenig" on the west side are also better seen in the stream (loc. 14) but exposures on the hillside should be examined for evidence of graded bedding indicating an upward succession from west to east. It should be noted that in the vicinity of Cnocan Donna the "Arenig" rocks have been indurated by proximity to the granite although the actual contact is not seen.

7. Note at this locality the altered and indurated gritty schists at the granite-schist contact.

From locality 7 to locality 8 continue westwards round the northern slopes of Creag Ghlas for about 1·6 km, noting various exposures of the coarse granite on the way. Then turn northwards up the headwater of the North Sannox which drains Coire nan Ceum to about the 335 m level.

8. At this locality a 2 m dyke of porphyritic felsite with pitchstone on its northern margin crosses the burn in a north-easterly direction. A short distance downstream is a thin northwest trending basaltic dyke. Note the fine series of moraines on the corrie floor a little higher up. These are among the most recent in the whole of Arran and were probably formed by a small glacier about 9500 B.C.

9. From locality 8 to locality 9 follow the burn downstream, noting on the way good sections of the coarse granite showing a rude platy structure due to the development of a system of close-set parallel joints. The granite is cut by a few basic dykes.

10. Examine here the nearly vertical junction of the granite against dark-blue phyllites with gritty bands, all much indurated.

11. In the North Sannox Water, just above the road bridge, there are excellent sections in steeply inclined greenish-grey schists and schistose grits, sometimes pebbly. Graded bedding, seen at a few places, again indicates an upward succession from west to east.

From locality 11 follow the Loch Ranza road westwards to locality 12. Note at the roadside, 500 m west of the bridge and near milestone 48, fine greenish-grey gritty schists, inclined steeply to east-northeast or northeast. A small quarry a little farther on shows a section in alternating fine grey grits and greenish slates all baked and indurated by proximity to the granite.

12. Examine carefully the vertical granite-schist boundary which follows the burn for some distance. Note the baked character of the schists, A 2·4 m, northeasterly directed, felsite dyke cuts across the junction. Note also the sections the burn has cut in the till. In the north of Arran the glacial till is grey in colour, reflecting the local predominance of granite and schist in contrast to the red sandstone-based till in the south of the island.

13. Here another granite-schist junction can be examined. The line of contact crosses a little stream in an irregular wavy line. The granite is fine-grained near the margin and contains some inclusions of schist.

Return along the road as far as the bridge over the North Sannox Water, noting on the way the striking and ever changing views of the granite mountains of north Arran.

14. The first exposures below the bridge are grey, gritty schists following those already seen at locality 11. Interbedded with these are fine-grained slaty schists and some 150 m downstream from the bridge comes the boundary between the Dalradian and the "Arenig". The actual junction can only be seen low down on the right bank, where the grey grits are brought against black phyllites

associated, a metre or so downstream, with a shattered or brecciated rock made up of fragments of much altered and carbonated lava. This probably represents a thin lava flow brecciated prior to final consolidation. The junction of the Dalradian and the "Arenig" should be carefully examined for signs of faulting (see p. 25).

The brecciated rock, exposed only on the south bank, is followed by a metre or so of black slaty schist with some thin irregular cherty layers and then by a thick bed of "greenstone", the lower lava of the "Arenig" succession, which extends downstream for about 90m. This is a fine- to medium-grained, greenish or greenish-grey rock and shows the characteristic "pillow structure" of lavas that have flowed into water. The chilled margins of the pillows (30cm or so in diameter) should be noted as well as the presence of numerous amygdales or steam-cavities filled, on their inner surfaces, with infiltrated mineral matter. This bed of lava is about 113m thick. It is followed downstream, first by a brecciated lava flow similar to that occurring near the base of the succession, then by a belt of thinly laminated black shales with pale-coloured cherty layers throughout and occasional thin ashy beds. This belt, about 30m thick, forms the main outcrop of the "Arenig" sediments. Anderson and Pringle (1944, p. 84) record the occurrence of a 3m wide intrusion of decomposed and carbonated gabbro which has invaded and indurated the shales. The intrusion can be seen on the south bank and in the bed of the stream. The thick upper lava which follows is only partly exposed here, but a good section of it can be examined on the hillside to the south at locality 6.

15. Examine here the section of steeply inclined grey gritty siliceous schists which strongly resemble those seen just below the road bridge near locality 11. There is no evidence in the stream itself of the relationship of these eastern schists to the "Arenig", but if the grits are followed up the hillside to the south there are indications of a faulted junction at one or two points (see loc. 6).

Graded bedding in the Dalradian outcrops west and east of the "Arenig" indicates in both cases an upward succession to the east.

16. At this locality the main band of black shales and cherts are exposed in the Allt Cairn Bhain, a little stream joining the North Sannox from the north. It was here that Anderson and Pringle obtained two small hingeless brachiopods referred to *Acrotreta* (see p. 25).

17. The junction of the Lower Old Red Sandstone with the Dalradian at this locality has been taken as a faulted one and is so shown on the 1:50,000 geological map of the island. This fault was regarded by Gunn (1903, p. 19), by Tyrrell (1928, p. 24) and Anderson (1945, p. 272) as a continuation of the Highland Boundary Fault. It is not seen in the stream but its position can be closely located. To the north it brings the Lower Old Red Sandstone successively against Dalradian, the upper lava of the "Arenig" and then the central belt of black shales.

Here purplish conglomerates are followed downstream by reddish sandstone and mudstone exposed at intervals and dipping east-northeast at angles varying from 20° to 40°. Southwest of North Sannox a well-marked band of conglomerate is exposed: this conglomerate can be followed from Glen Sannox to North Glen Sannox and beyond in an arcuate outcrop.

Note the remains of old dwellings along the north bank here. Bryce (1872) writes: "Many years ago, a large population, the largest then gathered in any one spot in Arran, inhabited this glen, and gained a scanty subsistence by fishing and by cultivating fertile plots on the sunny hillsides. In 1832 the whole of the families, amounting to 500 persons, were obliged to leave the island but were furnished with the means of reaching New Brunswick. They formed a settlement at Chaleur Bay, which became very prosperous. Garden enclosures, tree-clumps, solitary fairy thorns and ruined wallsteads, still remaining, give a meloncholy interest to this secluded glen."

From the North Sannox farm area, alternative routes are offered. For those interested in archaeology and the Upper Old Red Sandstone sediments underlying the lavas shown in fig. 11, the route lies generally northwestwards in Lower Old Red Sandstone strata. The alternative route misses out locality 19 and proceeds by way of the North Sannox picnic area from where the coast can be followed southwards towards locality 21.

18. Here a prominent feature formed by a massive conglomerate of Lower Old Red Sandstone age is surmounted by the remains of an old hill fort known as Torr an t'Sean Chaisteil. The plan of this structure (fig. 12) and the descriptive details that follow are taken from the *Book of Arran* published in 1910. The structure is there described as consisting of:

(*a*) an outer defence wall on the southeast, of which only the entrance or portal, about 1·2 m wide and small portions of the wall on either side remain.

(*b*) the main structure with an entrance 2·4 m wide and 3·7 m long leading into the central enclosure. The walls of this, probably originally 3·0 to 3·7 m in thickness, enclose an area about 37 m long and 29 m wide. On the western or northwestern side the wall seems to have been thicker because there is included in it a hut-dwelling, 2·4 m long and about 1·8 m wide, and the remains of what may have been two other similar chambers.

(*c*) an outer defence wall on the west side of the fort and at a lower level. This is now in a very ruined condition.

This fort certainly occupies a very commanding position with wide views of North Glen Sannox and the north Arran granite peaks. A traverse from this locality via the O.S. triangulation station (224 m) to the "Navigation Beacon" (009474) permits examination of the Upper Old Red Sandstone sediments which underlie a volcanic horizon mapped by Gunn. Fine features are formed of massive, whitish conglomerates which alternate with reddish flaggy sandstones with northeasterly dips.

19. At this locality (beside the beacon) evidence of the volcanic horizon should be sought. The lavas were described by Gunn (1903, p. 29) as purplish red, amygdaloidal and decomposed. The shore is reached by following the route of the overhead lines which carry power to the beacons.

20. In the vicinity of the lower "Navigation Beacon" cornstones occur in the sedimentary succession and evidence of faulting of

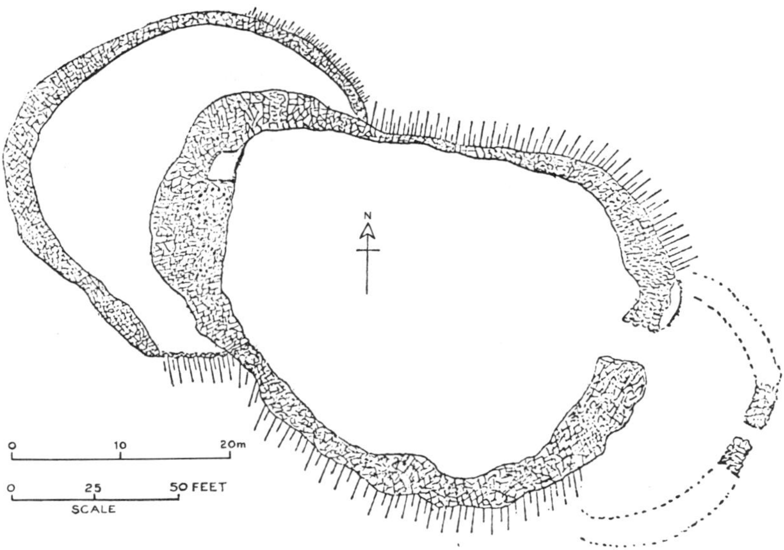

FIG. 12. Plan of the fort at Torr an t-Sean Chaisteil, Sannox; from *The Book of Arran*, 1910, p. 176.

various degrees of importance can be examined. On gaining the shore the excursion can be extended to the northwest, if time permits, to visit the land-slipped area of the Fallen Rocks where large masses, mainly of conglomerate, have fallen from the cliffs and obliterated the coastal shelf. It is just north of the Fallen Rocks that the basal Carboniferous beds appear.

21. Between the picnic area and Sannox Bay the Upper Old Red Sandstone sandstones and conglomerates now dip generally southeastwards. A conspicuous NNE trending composite dyke, with a width of some 16m, has caused obvious alteration of the sandstones. Nowhere is there any trace of the volcanic horizon, which might be expected to reappear at some point along this coastline.

22. Cross-bedded red pebbly sandstones occur here and are traversed by crush-zones and basic dykes. Note the large erratic block of granite (3 m by 2 m by 1·8 m) perched on coarse conglomerate and known as the Rocking Stone. Another granite erratic near it is 1·75 m high and 3 m long with a maximum width of 2·75 m. The sandstones exposed south of the Rocking Stone show on some of their surfaces a pattern of hardened cylindrical projections which Gunn suggested might be worm-casts.

Excursion 7: Loch Ranza, North Newton and Cock of Arran

(Figure 13, localities 1 to 17)

The objects of this excursion are:

(1) To continue the examination of the Dalradian schists, this time on the north margin of the north Arran granite mass;

(2) To examine the classic unconformity on the shore at North Newton;

(3) To study the Permian sandstones and breccias at the Cock of Arran and the land-slipped masses at An Scriodan.

1. Good exposures of dark-grey gritty schists, dipping northwards at steep angles (up to 50°) can be seen at this locality. Note that in passing along the shore towards locality 2 the northerly dip continues for a short distance while the schists become finer in grain and show much small-scale folding and contortion. The axis of the Catacol Synform is then crossed and the inclination of the schists is now in a generally southeasterly direction.

2. Here the steep southeasterly inclination is well seen in massive, greenish-grey, gritty schists with some darker slaty bands, all much folded. The rock-types closely resemble those exposed on the south shore of the loch near the pier. Boulders of a pale-grey sandstone, probably Carboniferous, are scattered along the shore towards Newton Point. Cowie (1905, pp. 157–158) has also recorded the occurrence at this locality of a “shoal” or band of rock fragments including types similar to those in the “cairn” at

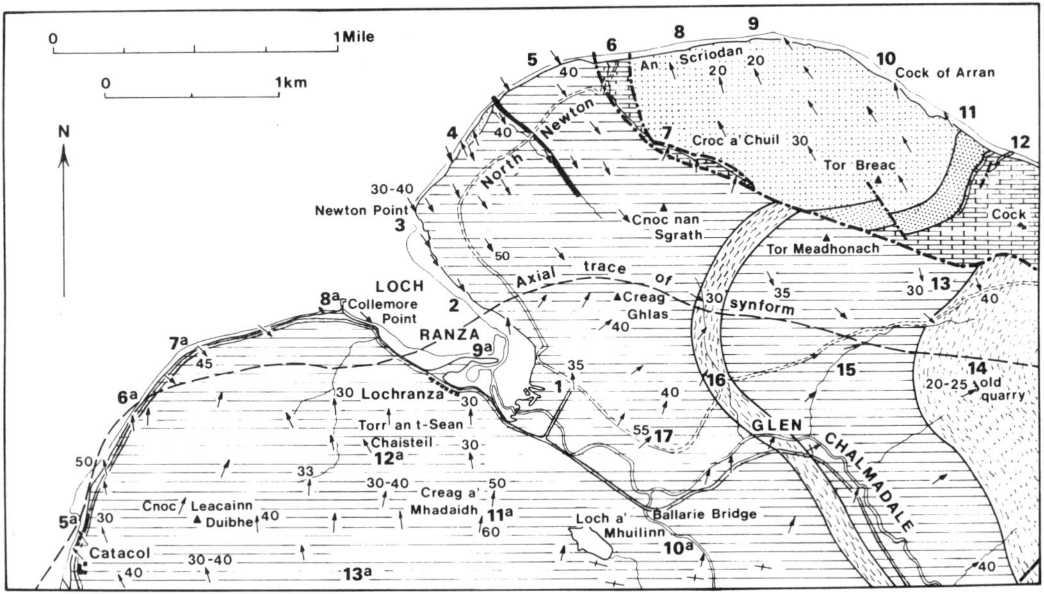

FIG. 13. Geological sketch-map of the Loch Ranza–Cock of Arran area to illustrate Excursions 8 and 9. For key see Fig. 5, p. 78.

Catacol, but without any of the red shaly crinoidal limestone obtained at the latter (see Excursion 8, loc. 5a).

3. At Newton Point some good examples of graded bedding can be studied in coarse or gritty beds, inclined to the south-southeast at angles of 30° to 40°. These are followed to the north by slaty micaceous schists, generally greenish or greenish-grey in colour.

4. Some 500 m northeast of the point, this is the locality of the well-known unconformity recognised by James Hutton as long ago as 1787. The section extends from the mouth of a little stream, the Allt Beithe, southwards along the coast for some 334 m. It shows reddish and yellowish sandstones associated with some bands of pale-coloured, sometimes nodular, cornstone resting with marked discordance upon the schists. These strata were at one time referred to the New Red Sandstone, but later assigned to the Calciferous Sandstone Series, the lowest subdivision of the Carboniferous succession (Gunn 1903, p. 52; Tyrrell 1928, p. 47). The lithological nature of the beds, however, suggests that they may be more fittingly placed in the upper part of the Upper Old Red Sandstone. Note that at the south end of the exposure they dip north-northwest at an angle of about 30°, while the underlying schists are inclined to the south-southwest at angles between 40° and 50°. The schists must have been nearly vertical at the time of deposition of the sandstones and cornstones.

The recognition of the North Newton unconformity provides a striking example of Hutton's genius and insight. It was described by him in Volume 1 of his *Theory of the Earth* (1795, p. 429) and additional details are contained in Volume 3 (published in 1899, pp. 235–236). The following quotation is taken from the latter: "Here the first thing that occurs is the immediate junction of the inclined strata of schistus and the other strata, which here appear to be a composition of sandstone and limestone; these strata are equally inclined with the schistus, but in the opposite direction. These two different kinds of stratified bodies rise to meet each other; they are somewhat confused at the immediate junction, but some of the sandstone or calcareous strata overlap the ends of the alpine schistus." Other unconformities recognised by Hutton in

Scotland are those in the River Jed, near Jedburgh, and at Siccar Point on the Berwickshire coast. Such examples of discordance in the geological succession were to him convincing evidence of cycles of change in the operations of nature, proofs of a "succession of worlds" following one another throughout geological time.

For an account of Hutton's life and work reference should be made to Volume 63, part 4, of the *Transactions of the Royal Society of Edinburgh* published in 1950 in commemoration of the 150th anniversary of his death.

5. Farther northeast along the coast towards the headland known as Rudha Creagan Dubh the principal rock-types are coarse and sometimes pebbly schists with prevailing southeasterly dips. The graded bedding seen on many of the rock faces shows, however, that the succession is inverted. A few northwesterly dykes are present.

6–7. Near locality 6 the schists end against a fault which brings on to the east strata belonging to the Carboniferous System. These Carboniferous rocks occur in a wedge-shaped strip, which extends up the Allt Mòr burn for about 1 km from the shore. The wedge is about 200 m wide on the coast but narrows when followed inland. Its eastern boundary is formed by another fault which brings on in turn Permian sandstones and breccias. The Carboniferous rocks exposed at intervals along the lower part of the burn are in the main white sandstones, but include also shales and, at one point, a reddish limestone which has yielded crinoid fragments and ribbed brachiopods (*Productus* etc.). This limestone is seen at locality 7, on the right bank of the burn. On the left bank, some 460 m above the mouth, there is a good example of fault-breccia.

Near the coast the Allt Mòr has formed a picturesque hollow, known as the Fairy Dell, cut in the relatively soft Carboniferous rocks and containing the remains of old dwellings.

8. To the east of the fault bounding the Carboniferous rocks the shore is occupied for some distance by Permian sandstones and conglomerates. Note that the generally north-northwesterly dip of these is approximately the same as the angle of slope of the hillside

above. The succession here consists of massive beds of bright red sandstone, cross-bedded and containing intercalations of conglomerate which become more numerous in the upper part. The sandstones and conglomerates closely resemble the Brodick Beds, with which they may be equated. The predominating pebbles in the conglomerates are schists of various types, quartzites and vein quartz.

9. This is the well-known land-slipped area of An Scriodan. It is made up of fallen or slipped masses of Permian sandstones and conglomerates which have covered the raised beach platform for a distance of some 640 m. These falls of rock, which are associated with deep rents and gashes in the hillside above, are said to have taken place over 200 years ago. Care should be taken in negotiating this part of the coast.

10. The name Cock of Arran is given to a large block of sandstone, some 6 m in length and 3 to 3·5 m in height, resting on the beach. Its upper part has been broken off, but originally it showed, when seen from the sea, a rather fanciful resemblance to a crowing cock. It lies about 27 m north-northwest of a spring near which some remains of old dwellings can be seen. Near the Cock itself there are a few dykes cutting the Permian: one of these, 1·2 m wide and running in a northeast-southwest direction, projects above the surface for most of its exposed course.

11–12. These localities are included in the excursion route to allow the visitor an opportunity to examine the contrasted types of sedimentary rocks encountered in crossing the Permian–Carboniferous boundary and of comparing these with the boundary succession of the same sequences at Corrie (Excursion 4, locs. 11 and 12). Towards the base of the Permian occur a number of conglomerate bands, two of which include pebbles of Carboniferous limestone, first discovered by Thomson and Wunsch in 1874 (Thomson and Wunsch 1877; Thomson 1897). The fossils obtained from these comprise corals, brachiopods, gastropods and crinoid columnals. The strata following the Permian to the southeast are assigned to the Coal Measures. Gunn records the occurrence of plant remains (referred to *Calamites*, *Cordaites*, *Mariopteris*,

Neuropteris and *Sphenophyllum*) in beds approximately 18m and 66m southeast of the boundary, and estimated the thickness of the Coal Measures here as about 76m. Leitch (1942, p. 149) gave a detailed section of the succession, with a thickness of 90·7m, the lowest beds being a 3·7m thick coarse white sandstone with a quartz conglomerate at the base. The varied sedimentation characters of these Coal Measures strata should be noted, characters which point to deposition under deltaic conditions. They differ from those on the Corrie shore in the prevalence of disturbed sandstones in which the bedding is often markedly twisted, contorted and overfolded, so much so as to suggest the formation of a sedimentary flow-breccia following deposition on a slope. Irregular bases to the different beds, minor unconformities and rapid lateral changes in facies are also common phenomena. These strata differ from the Corrie beds again in the general absence of bedded mudstones. Only one "mussel" has been recorded, a form which, according to Leitch, indicates a horizon low in the *modiolaris* zone or possibly in the *ovalis* zone.

If time permits the excursion may be extended about 1km to the southeast along the coast to look for the giant myriapod trail preserved in a sandstone bed in the Limestone Coal Group near Laggan (Briggs, Rolfe and Brannan 1979). A replica of the trail, along with a reconstruction of the 1m long millipede-like *Arthropleura,* which is thought to have made it, can be seen in the Arran Nature Centre at Brodick. The original trail, over 6m long and 36cm wide, runs east-west across the surface of a sandstone bed exposed in the old quarry above the remains of the salt-pans harbour northwest of Laggan (grid ref. 972511). **On no account should any attempt be made to collect material here as hammering would only lead to the destruction of this remarkable trace fossil.**

On the return journey to Loch Ranza follow the old track which runs from Cock Farm towards locality 13, noting that it crosses an important line of faulting which separates the Carboniferous sediments from the Dalradian schists.

13. Near this locality there are scattered outcrops of pebbly grits inclined to the south-southeast at angles of 30° to 35°. Cross

the hillside to locality 14. Between these localities the grits become less coarse and show a transitional junction with the Loch Ranza slate group. A similar interbedded junction occurs between localities 14 and 15. The exposures seen at localities 13 and 15 fall into subdivision A of figure 4 (p. 23).

14. Here there are two old quarries in beds belonging to the Loch Ranza slate group. Note that the inclination is now east-northeastwards at about 25°. This change in direction of dip, seen between locality 13, and localities 14 and 15, is due to the influence of the Catacol Synform. The rocks exposed in the old quarries are thin-bedded, mainly fine-grained slaty beds showing colour banding from pale greenish-grey to dark-grey. Some bands, however, are rough and gritty. The slates were worked on a small scale towards the end of the eighteenth century for local use at North Newton and Loch Ranza.

15. Examine sections of the Allt Chailean and beside the track showing strongly cleaved schistose grits inclined generally at 16° to 20° in an east-northeasterly direction.

16. The rocks at localities 16 and 17 fall into subdivisions B and C respectively of the inverted succession shown in figure 4 (p. 23). At locality 16 the schists show greenish-grey grits with intercalated slaty bands inclined to the northeast. Good sections are available in the little burn joining Glen Chalmadale from the north, 650m northeast of Ballarie Bridge, as well as in the main stream itself.

17. There are a number of exposures here alongside and above the old track showing banded greenish-grey and grey, gritty and often coarse schists, inclined to the northeast at generally steep angles, up to 60°.

Excursion 8: **Catacol, Loch Ranza and Creagan nan Caorach**

(Figure 14, localities 1a to 4a and 14a to 16a)
(Figure 13, localities 5a to 13a)

This excursion continues the examination of the Dalradian schists, this time in the area south-west of Loch Ranza, and the further evidence they provide for the basin-shaped structure known as the Catacol Synform. The axis of this synform is indicated on figures 13 and 14, and it will be seen that while northwesterly dips prevail on its eastern or southeastern side, the dips on the western or northwestern side are steeply inclined to the southeast. Opportunities are also given on this excursion for examining some striking exposures of the junction of the schists with the Northern Granite.

The visitor should, in the first place, examine the shore-section (fig. 14, locs. 1a–4a) to the southwest of Catacol Bay. The schists here are inclined in a general west-northwest direction at angles which range from 20° up to 60° or more. They are mostly hard, greenish-grey, gritty schists, with some bands of fine-grained bluish slaty schist. A good many quartz veins are present and a few thin basaltic dykes.

1a. Near this locality the axis of the Catacol Synform crosses the coast and passes out to sea. Much small-scale folding and crumpling is a feature of the schists at this point. Note carefully the different inclination of the schists on the two sides of the axis.

2a. Good examples of inverted bedding can be examined along this part of the coast.

3a. Note the occurrence of numerous crush-lines or small faults traversing the greenish, fine-grained chloritic schists at this locality in a northwesterly direction. The low raised beach platform with the old line of sea-worn cliffs behind is a striking feature here.

4a. Catacol Bay. Note the low level, recent, alluvial terraces of the Catacol Burn and the Abhainn Bheag, and the remains of a higher terrace thought to have been an outwash fan deposited by meltwater from the glacier which occupied Glen Catacol in Lateglacial times. Note also the shingle bar along the shore.

The coping of the walls at the roadside north of the bridge over the Catacol Burn, and elsewhere, contains a number of blocks of white sandstone and reddish fossiliferous limestone of Carboniferous age (see below).

5a. Just north of Catacol the shore section shows mainly hard gritty schists with some coarser bands, often sharply folded. The axis of the Catacol Synform is here close at hand.

About 1·2km north of Catacol there is a bank of loose boulders only exposed at low water and separated from the beach by a shallow channel. Cowie described this in 1901 (Cowie 1905, pp. 156–158, and plate IV, fig. 1) as a shoal of blocks and shingle consisting mainly of fragments of a massive red limestone with *Productus* and other shells, pieces of red shaly limestone packed with shell debris (crinoids and polyzoa) together with pieces of red and grey sandstone, quartzite, conglomerate and dolerites of all sizes and generally angular. He suggested that the "shoal" forms part of a moraine lateral to ice moving down Kilbrannan Sound from the north. In the same paper he describes a shoal occurring on the north shore at Loch Ranza and containing in addition to pieces of granite and slate, boulders of limestone, sandstone and breccia similar to those at Catacol (Excursion 7, loc. 2).

Gunn (1903, p. 52, and in Tyrrell 1928, pp. 69–70) described this bank or shoal under the name of "Catacol Cairn", and gave a list of the fossils obtained from the loose limestone blocks. This list includes corals, brachiopods, lamellibranchs and crinoid remains. He rejected the idea of a moraine as not very probable since the "nearest place where rocks of this kind occur *in situ* is more than 2 miles away to the east of North Newton", and suggested that the blocks may have been cast up by storms from an outcrop not far away, beneath the sea. B.N.Peach (in Gunn 1903, p. 138) accepted this view.

The reader should note that Gunn makes no mention of the similar "shoal" on the north side of Loch Ranza. Perhaps the source of the Carboniferous rocks of both localities is an undersea extension of the outcrops in the Allt Mòr (Excursion 7, locs. 6 and 7), a view which appears to be supported by geophysical evidence (McLean and Deegan 1978, p. 97) for rocks of Lower Carboniferous and New Red Sandstone age, not far offshore.

6a. Here the axis of the Catacol Synform again crosses the coast at a point close to the house known as Achnamara and a little over 1·2km from Loch Ranza Pier. Between localities 5a and 6a the shore-section shows hard, massive grits dipping a little west of north at angles of 45° up to 70°. Graded bedding, seen at a few points, again indicates inversion. Note the old sea cliffs and caves at the back of the Main Postglacial shoreline.

7a–8a. The axis of the synform has now been crossed and the inclination of the schists between here and the pier at Loch Ranza is to the southeast. The rocks are, in the main, coarse schistose-grits; the larger pebbles, occasionally up to 2cm in diameter, consist mainly of white quartz and are set in a finer-grained matrix of more or less crushed material. Graded bedding, showing inversion, occurs. A few crush-lines and some thin dykes traverse the schists; note the branching dyke a little west of the pier.

9a. Loch Ranza is a fjord-like inlet about 500m wide, which receives the united drainage of Glen Chalmadale and Gleann Easan Biorach. Loch Ranza Castle stands at the end of a sand and gravel spit which projects from the southwest side of the loch. This spit probably represents a portion of the old terrace of Postglacial times which has been cut up and largely eroded. The castle itself consists of an oblong main structure, 20m long and 10·7m wide, and has a small square tower projecting from the southwest corner. It was a three-storied building and originally the northeast corner was also carried up as a tower; this part, however, fell during a storm about 1897. Little is known of its history. The present structure, assigned to the latter part of the 16th century, occupies the site of earlier buildings (see Whitelaw 1910, pp. 246-249, fig. 2 and plates XLIII–XLIV).

Follow the road through Loch Ranza village to the junction of the Easan Biorach with the Chalmadale Water at Ballarie Bridge. A good section of sands and gravels belonging to one of the higher raised beaches is exposed above the junction and west of Ballarie. From a gravel pit here Cowie (1905, p. 158) obtained fragments of limestone with *Productus* and plant remains.

On a piece of level ground on the north side of the Chalmadale and almost opposite Ballarie Farm was the site of the chapel of St. Bride. In the *Lord of the Isles* Sir Walter Scott makes it the site of a convent in which the Maid of Lorne lived for a time, but while hardly a trace of foundations now remains, there is no historic evidence to show that anything more than a simple chapel existed.

10a. Ascend the Easan Biorach from the main road to this locality. The discordance in level between the Easan Biorach and the Chalmadale was explained by Gunn (1903, p. 14) as due to the presence of a belt of indurated and altered schist in which the tributary burn has cut a narrow gorge. This view was accepted by Scott (1918, p. 97) rather than the interpretation of Mort (1914) who attributed the discordance to glacial overdeepening of Glen Chalmadale (see p. 58).

11a. On Creag a'Mhadaidh (*circa* 230 m) examine the massive, pebbly schistose grits exposed at a number of places, and note the generally northerly inclinations at high angles.

12a. It is suggested that the route taken from locality 11a should follow the strike of the schists in the direction of Torr an t-Sean Chaisteil (*circa* 180 m). This is an old oblong enclosure some 70 m long and 36 m wide, with roughly constructed walls. It probably served as a shelter for cattle in times of raids, rather than as a fort. The rocks in the vicinity are mainly coarse-grained siliceous schists with pebbles of quartz or of quartz and feldspar up to 13 mm in diameter.

13a. From locality 12a ascend the hillside to locality 13a and thence to the granite-schist boundary at Creagan nan Caoroch (284 m) just beyond the southern margin of figure 13. During this traverse, at a level of between 240 and 290 m, a broad plateau-like feature is crossed, its surface largely covered by peat with exposures of rock to be seen here and there. This plateau forms part of the 300 m (1000 foot) Platform, a striking feature in the topography of the ground around the Arran mountains (see p. 57). It is well seen, for example, at the 230 to 300 m level, above Pirnmill and Thundergay (Excursion 10).

As the granite margin is approached the gritty or pebbly schists become indurated and the finer slaty bands show intense crumpling and puckering, while the inclination steepens until in places it is almost vertical. This belt of indurated schist extends westwards to a distance of some 225 or 275 m from the granite and forms a striking ridge-like feature on the hillside, which is actually higher than the adjoining granite.

14a–15a. Follow the granite-schist boundary in a west-southwest direction towards locality 14a and the head of Abhainn Bheag (fig. 14) noting the sharply defined nature of the boundary and the fact that the granite is generally fine-grained where seen in close contact with the schist. Good junctions are seen at a point a little west of the Abhainn Bheag and again, farther west, at locality 15a. The latter is the well-known Madadh Lounie locality where the granite cuts across the schists in a cliff face.

16a. Good exposures can be obtained here of the belt of fine-grained, blue-coloured slaty rocks which form a definite horizon in the schist succession at this locality and farther south. The belt is shown in figure 14. On its northern side at this locality it is followed by a band of coarse pebbly grit but it should be remembered that the whole schist succession traversed on this excursion is an inverted one and that while the pebbly grit *structurally* overlies the slaty rocks it underlies the latter *stratigraphically*.

On the way back to Loch Ranza by Cnoc Leacainn Duibhe examine sections in the schists for evidence of graded bedding showing inversion. The descent to the village may be made through the woods above Loch Ranza Pier (Coillemore), where the remains of an old bloomery can be seen. These bloomeries indicate the site of simple, primitive furnaces, in which the making of iron from bog-iron ore (or "bog-mine" as it was called) smelted by wood charcoal, was at one time carried out. Evidence of such small-scale smelting is widespread in various parts of the Highlands. The slag found at such sites is generally of a dense black iron-rich type as a result of imperfect methods of smelting in vogue. The name "bloom" was given to the lump or mass of iron produced.

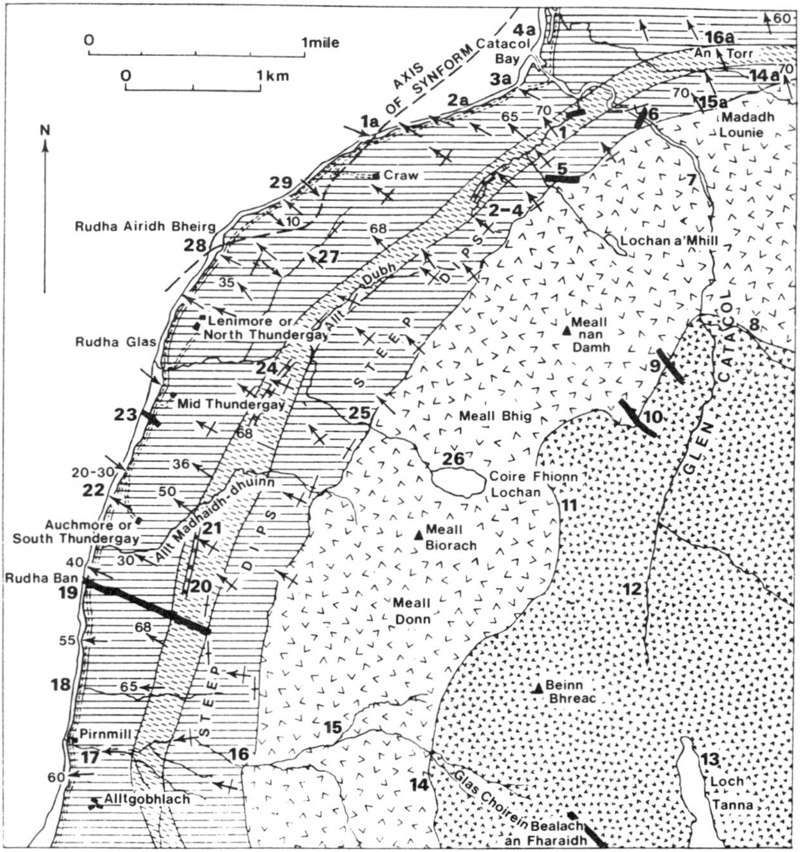

FIG. 14. Geological sketch-map of the Catacol–Pirnmill area to illustrate Excursions 9 and 10. For key to map, see p. 78.

Excursion 9: **Catacol, Loch Tanna, Pirnmill**

(Figure 14, localities 1 to 17)

This excursion takes the visitor up Glen Catacol to Loch Tanna and thence by a traverse westwards across the granite ridge to Glas Choirein and the coast at Pirnmill. The main purposes of the excursion are to examine (1) sections in the steeply inclined Dalradian schists along the western edge of the Northern Granite; (2) contacts of the outer, coarse grained, granite and the inner, fine grained, granite exposed in Glen Catacol and on the high ground to the west. The two types of granite contrast not only in their texture but also in their joint systems and mode of weathering. The wide-spaced mural jointing in the coarse variety and the more compact close-set jointing in the fine are reflected in the topography.

1. To reach this locality ascend the little stream which joins the Catacol from the southwest at the first sharp bend above its mouth. The stream, the Allt nan Eireannach (the Irishman's Burn), for some distance follows a belt of "dark slaty schists" which Gunn (1903, p. 14) describes as "probably identical with that which appears near the foot of Abhainn Beag on the east side of Glen Catacol". He adds that "there is a strong probability that the same band continues southwards from Glen Catacol nearly parallel to the boundary of the granite, but at no great distance, as far as the Allt Gobhlach opposite Pirnmill". The sections of these rather dark, finely-schistose rocks exposed at locality 1 and upstream should be carefully examined and compared with the coarse gritty schists on either side. Boundaries, however, cannot be sharply defined.

A short distance upstream from the High Lateglacial Raised Beach terrace a dyke of spessartite, 2m wide, crosses the burn. A hand lens shows dark prisms of hornblende in a pink feldspathic base. Unlike the dykes to be seen at localities 5 and 6, this spessartite has scarcely been affected by the intrusion of the granite. Careful examination will show the presence of yellow epidote.

2. Here steeply dipping grey or greenish-grey grits are followed westwards by the slaty beds of locality 1 and a traverse should be made across these to look for any evidence of transitional types. All the schists along the granite margin here form part of the upturned limb of the Catacol Synform produced during the emplacement of the granite (see p. 24).

3, 4 and 5. At these three localities the junction of the granite against the steeply inclined schists can be examined. Note that at locality 4 the latter are cut by veins of granite. The induration of the schists extends outwards for a distance of some 180m.

At locality 5 a dyke of diorite, some 3m wide, occurs. At its eastern end it is cut off by the granite and traversed by narrow veins of aplite. It will be noted that the dark minerals of the diorite are dull, resulting from the replacement of the original amphibole and pyroxene by little crystals of randomly orientated biotite at the time of emplacement of the granite.

6. Here the Catacol has cut a gorge through the indurated schists at the contact, the line of which is continuous with the section exposed on the crags to the northwest at Madadh Lounie (Excursion 8, loc. 15a).

The thermal metamorphic alteration of the schists has produced the minerals cordierite, biotite and corundum. A dyke of diorite, rather similar to that seen at locality 5, cuts the altered schists towards the upstream end of the gorge: it is about 5m wide and runs in a north-northwest direction. It is seen again on the rough track which follows the bank of the Catacol Burn. The dyke rock exposed in the burn was only mildly affected by the intrusion of the granite and shows glistening prisms of amphibole and equant crystals of augite in a base of white feldspar. The rock seen on the track is metamorphosed but to a lesser degree than that suffered by the intrusion at locality 5. An unaltered dyke of tholeiite, with phenocrysts of feldspar and xenocrysts of feldspar and quartz, is exposed crossing the Catacol in a north-northwest direction just downstream from the diorite. This dyke is clearly younger than the granite.

7. Ascend Glen Catacol to locality 7, examining sections in the

coarse granite at various points along the stream. On the hillside which rises steeply on the east to Creag na h'Iolaire there are numerous exposures illustrating its variations in texture, its joint system and its rough craggy mode of weathering.

8. Here the contacts of the fine and coarse granites should be looked for in the Allt Diomhan, where, however, it is somewhat obscured by crushing, and again in the Catacol a little below the confluence of the two streams. Flett (1942, p. 188) notes that at the latter locality the contact is well seen in a dry gully where the outward-dipping relationship of the fine to coarse granite "is clearly shown by the V-ing of the junction line towards the coarse".

Note that between the lower and upper portions of the Diomhan glen there is a series of cascades over which the burn falls 60 m to its junction with the Catacol. This discordance in level was interpreted by Mort (1914, p. 400) as evidence of glacial overdeepening of the latter. Scott (1918, p. 96) and Gregory (1920, p. 157) attribute it, on the other hand, to differential stream erosion, the older and larger Catacol following a zone of fractured and shattered granite.

9, 10 and 11. Follow the junction line to localities 9 and 10, at both of which it is cut by basic dykes. Between 10 and 11 its position cannot be clearly defined owing to lack of exposure but it would appear to show an embayment to the west. At locality 11 the inner granite is very fine-grained against the coarse and here again the evidence suggests a steep inclination towards the coarse outer granite. The visitor should note that in places veins of fine texture in the coarse granite sometimes stand out as ridges as a result of differential weathering.

12. From locality 11 cross to locality 12 at the head of the Catacol and thence continue southwards to Loch Tanna. This part of the route lies wholly within the finer-grained inner granite.

13. Loch Tanna, about 1·6 km in length, is the largest fresh-water loch on the island. It lies in a peaty drift-covered hollow at an elevation of 320 m and drains southwards towards the Iorsa Water. The route now lies westwards across the granite hill by the pass

known as Bealach an Fharaidh, between Beinn Bhreac and Beinn Bharrain. This traverse takes the visitor across both varieties of the granite and provides many opportunities for examining and comparing them.

14. In the long wide hollow known as Glas Choirein the contact can be seen at locality 14 where it is sharply defined in places and where the evidence suggests a steep inclination of the fine towards the coarse.

15–16. The route between these localities crosses the outcrop of the coarse granite (locality 15) and at locality 16 the intrusive junction of the granite with steeply dipping, indurated Dalradian schists, is seen. It should be noted also that between these localities the 300 m (1000 foot) Platform has been crossed, roughly between the 275 m and 290 m levels. Some 600 m up from the mouth of the Allt Gobhlach the stream cuts through a zone of fine-grained greenish-grey mica schists with thin dark slaty bands corresponding in position to the belt of similar rocks encountered at locality 1 (see also Excursion 10, locs. 20, 21 and 24).

17. Here there are good sections in steeply inclined (up to 56°) grey gritty schists.

Excursion 10: **Pirnmill to Catacol**

(Figure 14, localities 18 to 29)

This excursion has for its main objects the study of the rock-types and structures shown by the Dalradian schists along the western margin of the Northern Granite. In traversing the suggested route attention should be directed in particular to the following features:

(*a*) The rocks to be examined are for the most part altered schistose grits, representing very ancient coarse-grained arenaceous (sandy) sediments. Associated with these are a number of finer-grained phyllites, representing metamorphosed mudstones or sandy mudstones.

(*b*) A mappable belt or zone of dark phyllites and fine-grained grits, associated at localities 21 and 24 with a thin band of impure schistose limestone, can be traced running in a general south-southwest direction midway between the granite margin and the coast.

(*c*) The Dalradian schists along this tract lie almost entirely on the eastern flank of the Catacol Synform (see p. 24), the axis of which crosses the coast at localities 28 and 1a on figure 14.

(*d*) Graded bedding should be carefully looked for in the various exposures. It indicates that the original succession of beds is inverted and that the schistose grits in the centre of the synform are *older* than the slaty rocks and grits rising from below them to the east. The implications of graded bedding and its bearing on the structural disposition of the Dalradian generally is discussed briefly on page 22.

(*e*) A low raised beach is present along most of the coast-section covered by this excursion. Fragments of higher beaches may be seen at localities 28 and 29.

18. Examine the coastal section at this locality carefully. The rock types seen are mainly hard gritty schists with some pebbly bands and a number of ramifying quartz-veins. They are all much cleaved and folded, but it is possible in places to recognise that the cleavage or foliation planes are inclined at low angles to the planes of bedding.

19. At the headland known as Rudha Ban and immediately to the south of it the rocks are again nearly all coarse and gritty greenish-grey schists. Some crumpled and distorted bands of fine-grained greenish mica-schist are also present as well as pebbly bands of quartz (mainly) up to 12 cm in length. Note that the true dip is generally west-northwest, while that of the foliation is more westerly. Sometimes the two coincide.

A 12 to 15 m dyke cuts the schists here, indurating them along both margins. It is a coarse-grained crinanite carrying a little analcime and can be traced to the east-southeast for about 1·2 km. It does not, however, cut the granite. Some 200 m south of the dyke there is a marked fault-line showing a crush-breccia made up of schist fragments. Other similar crush-lines occur along the coast here.

From locality 19 ascend the hillside to examine exposures in the belt of dark, slaty schists already referred to. The same slaty group of rocks can also be examined on excursion 8 (loc. 16a) and on excursion 9 (loc. 1).

20–21. The belt of dark, sometimes black, slaty rocks sometimes includes at these localities a thin bed of limestone, grey in colour with darker partings, inclined at steep angles to the west-northwest. It is poorly exposed now but was described by Gunn (1903, p. 14) as "a limestone 4 or 5 feet (1·2 or 1·5 m) thick, which was formerly quarried and burnt".

22. Strong, gritty, grey schists, often much folded and broken by small-scale faulting and lines of crush, are seen along this part of the shore. Some of the crush-lines are accompanied by marked breccias. Note also the local abundance of quartz-veins.

Close to where the track from Auchmore (South Thundergay) Farm reaches the shore, there are several basalt dykes cutting the schists, one of them about 18 m wide.

23. Between the last locality and locality 23 the coastal exposures show much folded grey and greenish-grey schists varying from coarse-grained to fine. Note the changing and sometimes undulating inclination of the foliation along this part of the coast, suggesting that the axis of the synform lies close to the land. Northwesterly basalt dykes traverse the schists at several points, indurating them along the lines of contact.

From locality 23 follow the burn rising in Coire an Lochan up to locality 24. The lower part of this burn is known as the Allt Mhòr.

24. Note here the transition from the coarse, gritty grey schists to the fine-grained, dark, slaty rocks of the central belt. It should also be noted that here again a thin band of impure limestone is present and that the slaty rocks show along their western edge a schistose conglomerate with quartz pebbles up to 10 or 12 cm in length.

25. At this locality examine the steeply inclined gritty schists indurated by proximity to the granite. Along the schist-granite

junction veins of granite in places penetrate the schist. Exposures of the granite in the burn above locality 25 should also be examined and the character of the rock noted.

26. Coire Fhionn Lochan is a rock basin lying at an elevation of 329 m and surrounded by a granite rampart formed by Meall Biorach (549 m) to the south, the ridge of high ground overlooking upper Glen Catacol to the east and Meall Bhig (435 m) to the northeast. The loch, with a little beach of granite sand at its outlet, is certainly the most picturesque in Arran.

27. On the hillside known as Cnoc Moine nan Cuile (194 m) examine the exposures of massive, grey, gritty and often pebbly schist. Here, and in neighbouring outcrops the foliation dip varies a good deal in direction and amount. This locality is not far from the axis of the Catacol Synform.

28. South of this locality as far as Rudha Glas the coast shows good sections in much folded and faulted greenish-grey mica-schists often carrying veins and nodular masses of quartz. At locality 28 itself the axis of the synform crosses the coast and runs inland for about 800 m before reaching the shore again at locality 1a (see Excursion 8).

29. Between localities 28 and 29, the grey and greenish-grey gritty schists all lie close to the axis of the synform and on its western flank. Dips are now generally southeasterly but there is much folding and contortion.

Excursion 11a: Kingscross Point, Whiting Bay, Glen Ashdale

(Figure 15, localities 1 to 15)

The objects of this excursion are to examine phenomena connected with the intrusion of the Kingscross "crinanite sheet", to study the New Red Sandstone sediments and the dykes intruded into them, and to visit composite sills of the quartz-dolerite-felsite suite.

The excursion is set out assuming a start is made at Kingscross Point, but if the tides are not favourable the exposures in Glen Ashdale may be visited first. If an approach is made from Lamlash, Kingscross Point is reached by following the road which passes Kingscross Post Office. From Whiting Bay the main road is followed to Sandbraes from where paths lead to the Point.

1. Here the analcime-olivine-dolerite or gabbro ("crinanite" of Tyrrell) of the Kingscross sheet and its relations to the New Red Sandstone sediments can be examined. Tyrrell (1928, pp. 116–117) gives a description of the exposures around Kingscross Point, together with a section of the intrusion and sketches of basaltic veins in the "crinanite". More recently the locality has been redescribed by Tomkeieff and Longstaff (1961). From Kingscross Point westwards the following exposures may be visited. Just west of the Point a dipping surface of coarse dolerite vanishes under shingle which is almost certainly underlain by sandstone. Some 180m farther west baked sandstone is seen to dip under and to have an intricate contact with the chilled margin of a leaf of the Kingscross intrusion, the lower part of which is relatively coarse. "Veining" of the members of the leaf by very coarse grained dolerite and by dark basalt occurs. In the vicinity of a boathouse 450m west of the Point another leaf is exposed but not on the shore. The evidence points to all leaves of the Kingscross intrusion dipping generally to the northwest at moderate angles.

2. South of the point and just east of the ruins of "pill-boxes" the veined dolerite referred to above should be examined closely. It will be seen that the dolerite encloses stratiform layers of a dark basalt which develop tachylitic margins against the dolerite. These layers dip generally northwestwards, i.e. in the same direction as the leaves of the intrusion already mentioned. This rock with its tachylitic edges was regarded by Tyrrell as intrusive into the crinanite and referred to the quartz-dolerite suite. Tomkeieff and Longstaff, on the other hand, considered the magmatic history to have been (*op. cit.*, p. 201):

1. The intrusion of a sheet of olivine-dolerite which passed towards the margins into olivine-basalt;

2. The intrusion into this sheet of a crinanite magma accompanied by
3. The fragmentation of the olivine-dolerite into the slab-like and shred-like fragments, which are interpreted as inclusions in the crinanite;
4. Mobilisation of the melted basalt and intrusion of the rheomorphic basalt into the adjacent crinanite, dolerite or basalt.

The base of the intrusion is seen at many points along the shore between 90 and 365 m south of Kingscross Point. The igneous rock chills against, and includes rafts and fragments of sediment (often intensely baked) and shows intricate junctions. The most southerly analcime-dolerite seen on the shore is overlain by a thin sill of quartz-dolerite.

3. At this point is a pit in reddish, raised beach and deltaic, sands and gravels.

4. Along the main road between its junction with the road to Knockankelly and Kingscross Bridge are to be seen:

(*a*) Coarse analcime-dolerite enclosing at the south end small rafts of sandstone, and at its northern end showing a chilled surface dipping to the northwest under red drift.

(*b*) After a stretch showing only drift deposits a dyke-like body of felsite rises through sandstone and connects with a composite felsite-quartz-dolerite sill. Locally this rests on and appears to cut an intrusion of altered olivine-dolerite.

(*c*) A short distance south of Kingscross Bridge a typical coarse-grained analcime dolerite is seen again just above road level. This rock and that at 4(*a*) are parts of Tyrrell's Kingscross "crinanite" intrusion.

5. At this locality soft red sandstones, lying almost horizontally and showing good ripple-marks, are cut by numerous basic dykes. The latter have been examined by Rao (1959, pp. 240 *et. seq.*). He found it possible to determine the order of intrusion from the evidence afforded by intersections exposed at suitable states of the tide. The longest dyke, striking NNW–SSE, is composed of crinanite. It cuts three tholeiite dykes which strike NW–SE and

WNW–ESE. As the commoner relationship is tholeiite cutting crinanite this occurrence suggests that crinanite magma was available for intrusion on at least two occasions.

6. Here the more usual relationship between crinanite and tholeiite dykes may be observed. Near low-water mark a NW–SE crinanite dyke is cut by a thin NNE–SSW dyke of olivine-tholeiite (Brunton type). Rao (*op. cit.*) notes that the tholeiite dyke has developed tachylitic margins against the crinanite.

Glen Ashdale

7. This locality is reached by the road which leaves the main road just south of the Whiting Bay Hotel and continues past the houses in the glen as a forestry track. Near the gate at the start of the forest a stream falls over the lower part of the Glen Ashdale quartz-dolerite sill. As the road is followed upwards, higher levels of the sill are exposed. At a little picnic area glacially smoothed dolerite is seen. The stream here, if followed upwards, exposes the upper felsitic facies of the sill, above which are sandy sediments which pass upwards into argillaceous (clay-rich) beds. Tyrrell (1928, p. 88) refers to this section as providing the clearest evidence that there is no stratigraphical break between the dominantly sandy lower divisions and the finer-grained upper divisions of the New Red Sandstone in Arran. Return by the forestry road towards the picnic area and take the footpath leading to the Glenashdale falls.

8. The Glenashdale Burn provides a good section through the greater part of the Glenashdale composite sill. Tyrrell, who estimated its thickness at 64m, has given the following generalised section:

(1) Upper felsite, with intercalations of sediment;
(2) Feldspathic quartz-dolerite or craignurite;
(3) Main quartz-dolerite;
(4) Lower felsite.

The lower felsite can be seen, with some difficulty, in a gully lying a short distance northeast of the waterfall. Here a subsidiary sill is separated from the basal member of the main sill by highly baked sandstone. The contact rock of the main sill is a banded quartz-felsite which grades upwards through granophyric felsite and feldspathic quartz-dolerite into normal dark quartz-dolerite. The face of the lower waterfall is formed by a dolerite dyke.

The main quartz-dolerite can be examined conveniently in the banks and bed of the stream for some distance upstream from the waterfall. It is a dark, medium-grained dolerite with specks of pyrites. Veins of dark basalt cut the dolerite in the bed of the stream southwest of the fall. A short distance downstream of a little waterfall, and some 37 m or so downstream from a stone wall, xenolithic dolerite appears. This occurrence has been interpreted by Tyrrell as indicating an interior contact. Upstream the quartz-dolerite is rather more feldspathic than normal, although dark and even basaltic varieties also occur.

9. About 90 m upstream from the wall the feldspathic quartz-dolerite is followed by a slabby-jointed, deeply weathered, highly feldspathic quartz-dolerite which is succeeded by spherulitic felsite. The top of the sill is seen at a bend in the burn 128 m upstream from the wall: it passes under baked sandstone above which comes a thin felsitic intrusion, the uppermost part of which is a dark-banded pitchstone-like rock.

10. The Baoileig composite sill is much thinner than the Glenashdale sill, its estimated maximum thickness being 12 m. The Allt Dhepin has cut a little gorge through the intrusion which can be readily seen to rest upon horizontal indurated sandstone and marl with a somewhat uneven base. The upper contact is seen at the top of the waterfall. The sediments into which this sill has been intruded are referred by the Geological Survey (IGS) to the lower part of the Auchenhew Beds.

The Baoileig sill has been examined by Rao who has given a diagrammatic vertical section (1958, p. 278) from which the following succession has been compiled:

	metres
1. Craignurite	3·0
2. Felsite	1·5
3. Craignurite	0·3

4.	Acidified quartz-dolerite	2·1
5.	Fine-grained quartz-dolerite	1·2
6.	Acidified quartz-dolerite	2·1
7.	Felsite	0·6
8.	Craignurite	1·2

The recognition of these rock types is a matter for laboratory rather than field study.

11. As an example of the felsite masses of the district which show no association with the more basic rock types, the Torr na Baoileig intrusion may be examined in the gorge here. Access is obtained by crossing the feature formed by the Baoileig quartz-dolerite a short distance east of the lower end of the gorge of the Baoileig Burn.

The rock is somewhat poorly preserved coarse spherulitic felsite or granophyre. Towards the south end of the gorge and on the left bank of the stream, the margin of the intrusion transgresses hardened sandstone and marl. Locally the contact rock is a pink and green banded felsite sometimes associated with a green pitchstone. The felsite and sediments are cut by a rotted basic dyke. In the face of the waterfall at the south end of the gorge the base of the Dippin "crinanite" sill can be seen to rest concordantly on the sediments. The marked feature formed by the Dippin sill should now be followed eastwards.

12. Here, as at other adjacent localities, it should be noted that the eroded upper surface of the alkali-dolerite stands higher than the sediments at its southern (upper) contact. A conspicuous rounded knoll lying just to the south is also of analcime-dolerite. Below the scarp of the Dippin intrusion and in the hollow between it and the eastern end of Torr na Baoileig (031244) is a WNW trending dyke of pitchstone. This dyke carries microphenocrysts of quartz, feldspar, soda-pyroxene and iron-rich olivine. Continue westwards towards locality 13 noting the continuous scarp of the Dippin sill and the higher sills of quartz-dolerite and felsite.

13. Just south of the prominent little felsite hill, Torr an Loisgte, a dyke of pitchstone associated with spherulitic felsite cuts

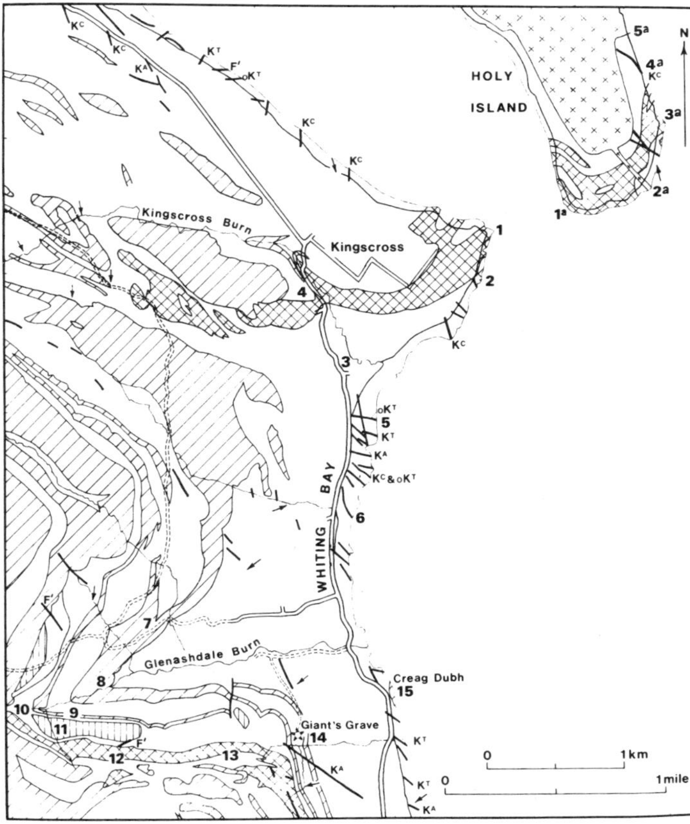

FIG. 15. Geological sketch-map of the Kingscross–Whiting Bay area to illustrate Excursions 11 and 11a. For key to ornament see Fig. 5, p. 78. Note that the Triassic sediments have been left blank. Where indicated the identity of the dykes is as follows: K^C crinanite, K^T tholeiite, $^OK^T$ olivine-tholeiite, K^A "andesitic" quartz-dolerite, F^I pitchstone.

the lower scarp of the Dippin sill. The latter intrusion also forms an upper feature cut by a pair of fresh tholeiite dykes and is overlain at its western end (039244) by a sill of vitreous quartz-dolerite. Downstream (041245) the lower leaf of the Dippin intrusion is cut through by a small stream giving rise to a steep cleft. Lower ground may be reached by following a track down the face of the escarpment a short distance east of the cleft.

14. The stream issuing from the cleft should be located and followed. It provides sections of the Baoileig and Glenashdale sills and associated sediments. The first sill forms blocky outcrops of feldspathic quartz-dolerite. The second is associated with sandstones which dip steeply westwards; the intrusion appears to dip similarly. It is composite with a lower dark dolerite component and an upper banded felsite.

The imposing chambered cairn known as the Giants' Graves lies between these sills and a little north of the stream. The main road is regained by the path which runs northwards from the "graves" to the bottom of the glen and then eastwards.

15. Creag Dhubh (Black Rock) is a good and readily accessible composite sill. Some 8m thick it is intruded into sandstones with which its dips at about 20° to the west-southwest. It has been described by Rao (1958, p. 278) and by Rogers and Gibson (1977, pp. 1–8). A median layer of quartz-hypersthene-dolerite is flanked symmetrically by felsites. The outer members are decomposed, buff-coloured, rust-spotted rocks with little patches of green chlorophaeite. The next rocks inwards are pale-grey feldspar-phyric felsites which have merging contacts with the median dolerite. The latter varies in coarseness and exhibits varying degrees of alteration probably due to hydrothermal activity at the time of intrusion.

Excursion 11b: **Holy Island**

(Figure 15, localities 1a to 5a)

Holy Island, dominating the approaches to Lamlash Bay and rising steeply to a height of 314m, is about 3km long and, at its widest part, 1 km broad. It derives its name from its association with St. Molaise or Molios, a disciple of Columba, who came here about A.D. 680. The island is mainly composed of a massive sill-like intrusion of trachyte. The thickness of this intrusion (240 m or so remain) and its failure to extend to the main island, suggests that it may be mushroom-shaped or laccolithic in form.

The island can be reached from Lamlash during the holiday season. Whilst waiting for the boat dykes on the shore may be examined, especially the composite dyke 180 m southwest of the site of the pier, and the NE–SW crinanite dyke 275 m northeast of the pier site. The position of the landing place on Holy Island, although necessitating a walk of about 1·6 km along the path which runs along the western side, will allow a visit to the cave or cell of St. Molaise. Runic inscriptions on the sandstone wall of the cave include one of 11th century date. There are six others from the 12th or 13th centuries one of which refers to King Haakon of Norway's expedition of 1263 (which culminated in his defeat at the Battle of Largs in that year). The cell was cleared of accumulated rock material in 1908 when a paved floor was revealed and a mass of "kitchen refuse", consisting of limpet and oyster shells and fragmentary bones, mainly of domesticated animals, was discovered. Nearby is a small well said to have been blessed by the saint and as a consequence the water is reputed to have curative properties.

1a. In the vicinity of the jetty at the lighthouse buildings coarse analcime-olivine-dolerite with pale-coloured veins is extensively exposed. Northwestwards from the buildings highly-baked sandstones outcrop; these rocks carry quartz paramorphs after tridymite. At the seaward end of this area of sediments the dolerite contact is seen to be intricate and transgressive. Farther to the north another strip of baked sandstone underlies a thin dolerite sill. There is little doubt that these intrusive rocks are the continuation of the Kingscross intrusion which, on Arran mainland, is also split by intercalations of sandstone.

2a. The foreshore and cliff at this point are cut by a 27 m thick dyke-like intrusion of trachyte with margins which locally show flow-banding. Although sufficiently fresh material for detailed examination has not been obtained, the rock closely resembles that of the laccolithic intrusion occupying the greater part of the island. The dyke may well be a feeder of the latter. Scree and downwash, however, cover its landward extension and the margin of the main intrusion is covered at the likely position of the junction.

3a. A short distance north of Pillar Rock Lighthouse (068292) the analcime-olivine-dolerite can again be seen. Here it encloses bands and patches of sandstone baked to hard quartzite in which once more tridymite has been developed. The junction between the dolerite and the enclosures is often intricate. The sill rock is generally quite fresh.

4a. The cliff backing the raised beach shows a shallow synclinal arrangement of well-bedded, bright red sandstones which exhibit carious weathering on a large scale. The overlying crags of trachyte show regular columnar structure. Trachyte debris forms the main component of a storm beach on this section of the coast. The sediments are cut by a NW trending dolerite dyke which can be traced up the lower cliff. It does not, however, cut the overlying trachyte.

5a. Here the margin of the trachyte turns abruptly east-northeast and cuts across the sandstones. Against the latter the igneous rock has a nearly vertical contact and it develops relatively small horizontal columns. A short distance to the north, normal vertical columns are displayed after a short stretch in which the jointing is poorly developed. In spite of its great thickness the intrusion has resulted in little alteration of the sediments, the red and brown colours of the latter being retained almost up to the contact.

The trachyte is a rather fine-grained rock which, when fresh, has a deep blue-grey colour. It breaks and rings like a phonolite giving rise to abundant slabby screes. It weathers readily to light-grey or buff colours. Rhythmic bands, due to weathering are often beautifully shown by this rock. These may be seen by fracturing weathered slabs. The fresh rock is interesting in that it contains a blue soda amphibole usually referred to riebeckite. Recent work has shown the mineral to be arfvedsonite. The amphibole in the Ailsa Craig micro-granite, the rock chemistry of which closely resembles that of the Holy Island, has recently been reclassified as a riebeckitic arfvedsonite (Howie *et al.* 1981, p. 126). The other mineral components of the Holy Island trachyte are sanidine, sodic ferrohedenbergite, iron ore and quartz.

Excursion 12: **Largybeg to Bennan**

(Figure 16, localities 1 to 16)

The object of this excursion is to examine the Triassic sediments of the southern part of Arran and the large number of Tertiary sills and dykes which occur there.

Approach Largybeg Point by the road leading from the main road at Largybeg, noting *en route* a good exposure of red and green sandy marls and sandstones dipping to the south-southwest. At the point where the road meets the cliff-top a thin sill of quartz-dolerite is seen.

1. At the northeast end the Largybeg Point section is terminated by a composite sheet which appears to dip steeply northeastwards and to cut transgressively across the sediments. The lower member of this intrusion is a light-coloured acid craignurite. This rock encloses elongated vesicles filled with a dark greenish substance which has been referred to chlorophaeite. The upper member, which is locally of considerable thickness, is a quartz-dolerite *aff.* Talaidh type. To the southwest lies the shore section described and figured by Tyrrell (1928, pp. 88–90 and 1915, p. 191). This section in descending order is as follows:

		metres
(*a*)	Thin-bedded red sandstone	12·8
(*b*)	Composite sill	4·6
(*c*)	White gritty sandstone, with line of local unconformity and rock-arch	10·0

(*d*)	Coarse grit with bands of conglomerate	2·7
(*e*)	Conglomerate	2·0
(*f*)	Carious white sandstone	2·7

These rocks dip at 15° or so to the southwest. They are cut by basaltic dykes, the one terminating the above section to the southwest forming a fine wall which stands up to 7·5 m above the shore.

The conglomerates are worthy of notice as they are of a type uncommon in the New Red Sandstone of Arran. As noted by Tyrrell the pebbles consist mostly of white and pink vein-quartz and quartzite with some of red feldspar porphyry, flinty quartz-felsite, grey mudstone, red sandstone and chert. He has recorded dreikanter which have suffered water erosion subsequent to wind-faceting and has noted that, while the igneous and quartz pebbles are generally well-rounded, the matrix is a coarse sandstone consisting mainly of rounded and polished grains of quartz with an occasional cleavage fragment of fresh feldspar embedded in a whitish decomposed feldspathic matrix. This suggested to him that the conglomerate represents an accumulation of pebbles and sand brought together by occasional rain bursts in a desert or semi-desert region. Many of the pebbles have a distinctly Lower Old Red Sandstone appearance.

The composite sill (*b*) has margins composed of acid craignurite and a centre of feldspathic dolerite. The marginal rocks enclose vesicles filled with an olive-green mineral (?chlorophaeite). In addition, the lower member holds vesicles filled with amethystine quartz. The dip of this intrusion is to the southwest, in conformity with the country rock. The two Standing Stones, erected on the landward extension of the sill, are composed of local pebbly sandstone.

2. On the foreshore a short distance southeast of the little cottage south of the Standing Stones two beds of sandstone contain numerous "nests" of clear crystalline calcite in cavities.

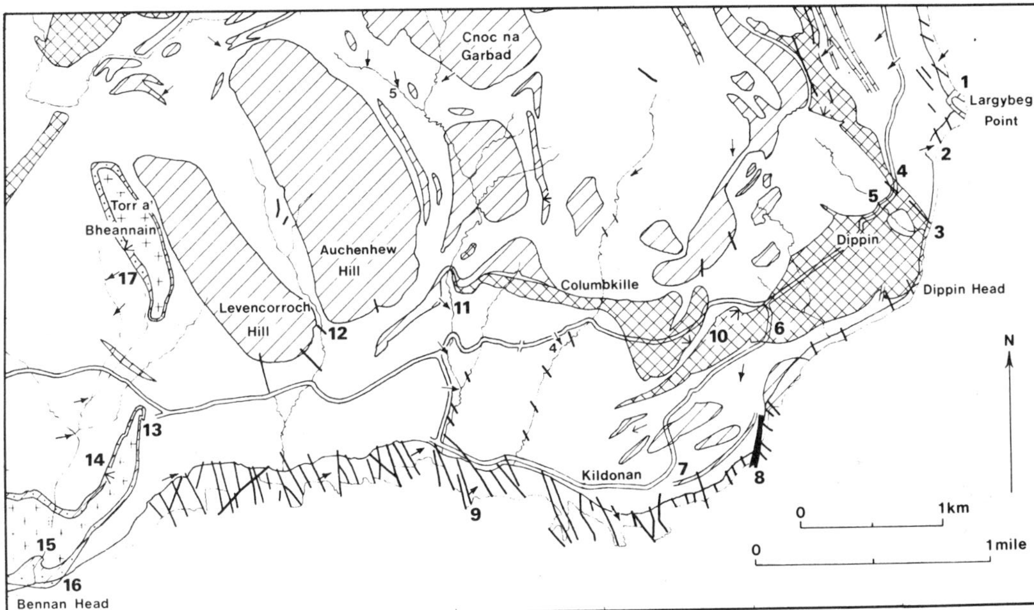

FIG. 16. Geological sketch-map of the Largybeg–Kildonan area to illustrate Excursion 12. For key to map see Fig. 5, p. 78. Note that the area occupied by Triassic sediments has been left blank.

These are of all shapes and sizes up to as much as 15 cm in diameter. The horizon of this locality is close to the junction between the Lamlash Beds and the Auchenhew Beds (p. 37).

Some 275 m southwest of Largybeg Point and near Port na Gaillin the foreshore is crossed by an irregular sheet of dolerite which bakes and disturbs the sediments. The latter lie just below the horizon taken as the base of the Auchenhew Beds. They consist of sandstone, generally red and soft, sometimes ripple-marked, and purplish shales or mudstones. The first dyke to the southwest of the dolerite sheet is split into four members by screens of sandstone baked to quartzite.

3. At this point the Dippin analcime-dolerite runs out to sea. A tholeiite dyke rises vertically through it, showing well-developed cross-jointing which contrasts strikingly with the vertical columnar jointing of the sill. The dyke shows good chilled margins. Just southwest of the angle of the cliff at the dyke, a mass of hard, purple, micaceous sandy mudstone is enclosed in the dolerite, with the development of the usual basaltic contact rocks. Northeast of the dyke and a little below high water mark a small patch of quartz-dolerite rests on sandstone with numerous ripple-marks. The attitude of the margin of the Dippin intrusion here is a matter for conjecture. Some 200 m to the south-southwest the old sea-cliff of analcime-dolerite is cut by a 1·6 m sill of typical quartz-dolerite.

Good fresh specimens of the Dippin sill are readily obtained. In hand specimen, plagioclase feldspar, augite and olivine can be distinguished. Thin sections show laths of plagioclase, ophitic purplish titanaugite, olivine which is generally partly serpentinised, iron ores and interstitial analcime. The main road should be regained by the road leading from the Trareoch Hotel. By the side of this road are seen outcrops of red sandy micaceous marls.

4. At this point the road crosses landslip material which has fallen from the inland cliff, An Garradh. On the west side of the road the slip has exposed sediments cut by a thin sill of quartz-dolerite. Below the sill the sediments are red, sandy, micaceous marls with green spots. Just above the sill similar red marls occur followed upwards by grey and red marls with bands of

calcareous nodules (potato stones). Against the marl the sill has developed irregular, chilled and banded margins. The sediments appear to be unaffected by the intrusion. This locality has yielded microfossils considered by Warrington (1973, pp. 109–116) to indicate a late Lower Trias or early Middle Trias age for the rocks.

5. Here two small quarries have been opened in the Dippin Sill. The rock is decidedly patchy. A darker dolerite encloses irregular masses of relatively light-coloured coarse-grained rock the latter being relatively rich in analcime (i.e. teschenitic). Locally spheroidal weathering is well marked. In both quarries the sill is cut by tholeiite dykes; these bear phenocrysts of calcic plagioclase, clino- and orthopyroxene, in a groundmass of these minerals; there is an abundant residuum of feldspathic material probably representing devitrified glass. On the southeast side of the more southerly quarry marginal rocks are seen in irregular but generally steeply inclined contact with the sediments.

6. On the way to Kildonan the road re-crosses the lower contact of the Dippin sill, the scarp edge of which is well marked running southwestwards parallel to and above the road. From this point on the road a good view of Pladda and, on clear days, Ailsa Craig may be obtained. The former island is largely formed of a sill of quartz-dolerite. The obvious low cliff is due to marine erosion in Main Postglacial Shoreline times.

7. Following the main road westwards from its junction with the road to Kildonan for 200m a strong feature is seen on the left. Examination of this shows that the upper margin of the Dippin intrusion (at grid ref. 039220) cannot be concordant as contact rocks are seen to dip steeply east-northeast. Similar rocks occur on the southeast-facing scarp which crosses the fields (once Kildonan Golf Course) farther west. The intrusion here is an irregular sheet of analcime-olivine-dolerite, which may be referred to as the Columbkille intrusion.

8. In the neighbourhood of Kildonan the High Lateglacial Raised Beach platform is well developed. The seaward margin of this feature is formed by a sill of craignurite felsite, the

flow-vesiculated central portion of which may be readily examined by the top of the steps which lead down to the shore a short distance east of the hotel. The Coastguard Station and the ruined Kildonan Castle both stand on this sill.

9. A traverse of the shore westwards towards Bennan Head will permit examination of Triassic sediments cut by a large number of basic dykes. The sediments are white and grey sandstones, often shaly and locally ripple-marked, with interbedded red, green and grey mudstones. They dip at low angles off an anticlinal axis which crosses the shore south of Levencorroch. The cliffs backing the Main Postglacial Shoreline give good sections of the sediments. In addition to the types already mentioned as occurring on the shore, bands of calcareous concretions and nodular, impure limestones are also to be seen.

The dykes are numerous; between the Allt Mòr and Port a'Ghille Glais some 60 dykes cross the foreshore. The predominant direction of these intrusions is NNW–SSE with a few trending NW–SE and NE–SW. Their width varies from 30 cm to 18 m or so. Dyke complexes and multiple dykes occur. Intersections can be studied under suitable tidal conditions. Bifurcation can also be seen; in fact most of the phenomena attendant on dyke intrusion can be studied on this shore. So far as is known the rock-types present belong to the crinanitic and tholeiitic suites.

9a. An alternative to the section west of the Kildonan Hotel is to examine the dykes on the shore to the east. Although the frequency of dykes is less here this locality is otherwise quite representative. Also it permits examination of the broad dyke complex which crosses the shore in a N–S direction 0·5 km northeast of the castle. This body, which consists mainly of dykes of coarse alkali-dolerite, is interesting in being cut by at least one tholeiite, and appearing to truncate NW–SE dykes which include both alkali-dolerites and olivine-tholeiites. On its western side and just above high water mark is a thin sill of tachylite.

10. Although awkward of access, the gorge at Allt Mòr, just below the waterfall Eas Mòr (grid ref. 020222), is worth a visit. The gorge terminates at the fall which lies at the centre of a fine

amphitheatre. The cliff on the western side of this feature shows the thinning Columbkille analcime-dolerite resting on light-coloured baked sandstone and enclosing long rafts of the same material. This intrusion is overlain by a paler felsite-quartz-dolerite sill which runs round the amphitheatre, forming the lip of the waterfall. The analcime-dolerite peters out in marls before it reaches the fall beneath which the sediments are cut by small faults and narrow dykes.

11. The Levencorroch Burn provides an excellent section in marly and sandy sediments forming part of the upper division of the Trias (Auchenhew Beds). On the east bank of the burn, 275 m upstream from the road-bridge, thick beds of marl with posts of sandstone outcrop. The marls weather to a characteristic cuboidal rubble. At the lower waterfall a thin basaltic sill is intruded into the sediments. Above this the stream flows in a small gorge the east wall of which shows marls with some sandstone, while the west wall shows preponderant sandstone. In the face of the waterfall at the head of the gorge a fault is clearly exposed. The above sediments are cut off upwards by the Levencorroch Hill–Auchenhew Hill quartz-dolerite sill which gives rise to the craggy scarps overlooking the main road.

12. At this point is an old quarry in the quartz-feldspar-porphyry which forms the central part of the Bennan composite intrusion. The rock, which is in the main far from fresh, is massive and carries obvious large phenocrysts of feldspar, and rather smaller phenocrysts of quartz in a yellow or brown fine-grained base. On the west side of the quarry there is a small exposure of basified, dark porphyry. The east side of the road leading to Craigdhu Farm should now be visited. In small quarries the tholeiitic upper member of the intrusion can be examined. It is a dark dolerite of medium grain. On close inspection occasional xenocrysts of quartz and feldspar can be seen. The margin is exposed in one small quarry where the basic rock is seen to dip steeply to the west through hardened gently-tilted sandstones.

In the field to the north and towards the main road there are exposures of porphyry and dolerite. The latter rock is seen here on the east side of the former. It is conjectured that the basic rock continues round the acid core on the north.

13. The west side of the ridge of Creag Dhubh should now be followed southwards and westwards towards the cairn on the west bank of the Struey Water. On the way the crags of dolerite above the farm should be noted. Their steep dip westwards is readily appreciated. On both banks of the Struey, dolerite is seen near the cairn. As usual it carries xenocrysts.

The section exposed in the Struey Water can be examined conveniently only if the water is low. In general the normal porphyry outcrops: there are, however, a few bands of darker, obviously more basic rock such as occur close to the upper and lower basic members of the intrusion.

14. Near a small waterfall at this locality dark, xenolithic porphyry outcrops. From this waterfall the burn flows in a gorge cut in the lower dolerite. At the seaward end occurs the Struey Falls. The shore may be reached by a path some 46m east of the Struey at the cliff top. Some scrambling is involved and extra care should be exercised.

15. A short distance east of the Struey Falls is Bennan Head and the Black Cave. At the latter large masses of sandstone, baked to quartzite, are isolated in the dolerite. Other patches of baked sediment are seen enclosed in whinstone near the foot of the Struey Falls. The lower dolerite, like the upper, is a dark rock of medium grain carrying xenocrysts of quartz and feldspar. In thin sections it is found to contain hypersthene.

16. A visit to Torr a'Bheannainn, lying north of the main road, will show that it is formed of quartz-feldspar-porphyry margined by quartz-dolerite with xenocrysts of quartz and feldspar. On the western side of the hill the dolerite dips at 50° to the southwest. Examination of this intrusion (Herriot 1972) has shown it to be related to the Creag Dhubh section of the Bennan composite intrusion.

Excursion 13: **Drumadoon and the Tormore Dykes**

(Figure 17, localities 1 to 9)

The object of this excursion is the examination of the composite dykes and other intrusions which abound on this stretch of shore. It can be carried out by public transport from the main centres on the east side of the island. If private transport is used it will be found convenient to arrange for collection at the appropriate terminal point, especially if another excursion is to be made on the same day.

The Tormore shore is reached by the track which runs westwards from the main road at the old post office (grid ref. 894325). The route follows the raised beach southwards past the houses.

1. The foreshore and raised beach cliff are eroded in sandstones which occupy horizons which are probably low in the Triassic succession of Arran. These sandstones are often strongly cross-laminated and careous weathering is locally striking. Irregular distribution of carbonate cementing material leads to hard, rough-weathering bands alternating with soft red sandstones and siltstones. Pseudomorphs after halite and dendritic gypsum are recognisable in places.

2. Along this stretch of shore numerous dykes outcrop including the composite dykes made famous by Judd (1893). The finer points of the petrography of these interesting intrusions can only be studied in the laboratory but the constituent rock-types are sufficiently diverse to enable adequate field determinations to be made. These types are pitchstone, felsite, quartz-feldspar-porphyry (quartz-porphyry of the 1:50,000 Geological Map) and dolerite (generally the tholeiitic variety). In the case of the dykes described by Judd the Roman numerals used by him have been appended to figure 17.

The most northerly pitchstone, although it enclosed areas of felsite, was not considered by Judd to be composite. Contrary to the information given in the Arran Memoir (Tyrrell 1928, fig. 30, p. 219) it forms a conspicuous feature on the foreshore. It is rather

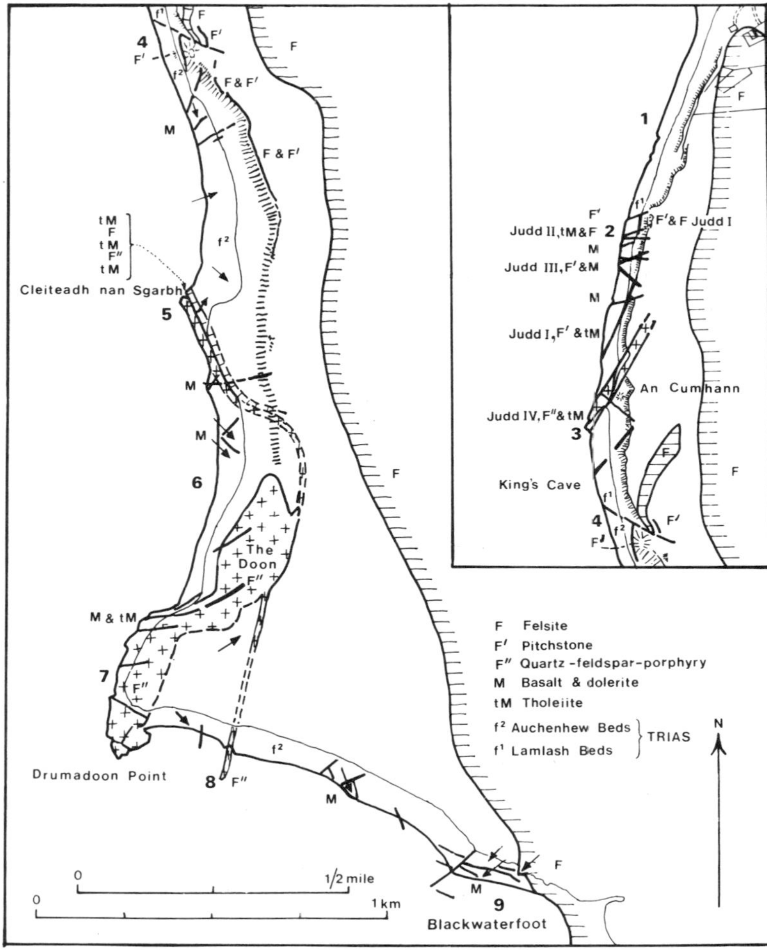

FIG. 17. Geological sketch-map of the Drumadoon–Tormore area to illustrate Excursion 13.

irregular in thickness and hade being some 5m thick at H.W.M., and it runs ENE–WSW. The predominant rock-type is a dark-green pitchstone.

Near H.W.M., a dyke (Judd I) of similar rock, about 4·6m thick and coursing generally N–S, can be examined readily. The mutual relations of these two intrusions are obscure.

Judd's dyke II consists of a central quartz-felsite 4·5m or so thick, bounded by spheroidally-weathering tholeiites 1·2 and 1·8m thick on the north and south margins respectively. Judd records the felsite passing in places into "pitchstone-porphyry or 'vitrophyric' rock" which is stated to occur "as a band varying in width from 15 to 60cm, sometimes forming part of the quartz-felsite mass and at other times intersecting the masses of andesite" (i.e. tholeiite).

Some 55m south of intrusion II two dykes appear to intersect: they have irregular courses before diverging as they cut the cliff.

A short distance farther south a NW–SE dyke (Judd's No. III) shows pitchstone 1·2 to 1·8m thick, on the north side of a thick dolerite. In this case the dolerite is an olivine-dolerite of a common Tertiary type and not of the tholeiitic type usually found in association with pitchstone and felsite. It is possible, therefore, that the association of rock types in this case is fortuitous and that the dyke is not really composite in the accepted sense.

South of dyke III and above H.W.M., the N–S trending pitchstone (Judd I) reappears. It now has a marked eastward hade. Southwards it swings round towards the south-southwest its dip gradually decreasing. By the development of felsic modifications and the presence of thin, generally rotted, marginal tholeiite, the intrusion becomes composite. The pitchstone is dark-green in colour; flow-banding is strikingly displayed, especially in dislodged blocks on the shore.

3. At An Cumhann a 27m composite dyke (No. IV) of quartz-feldspar-porphyry and tholeiite makes a strong feature. This body shows to advantage all the features of porphyry/tholeiite intrustions, i.e. the marginal tholeiites with xenocrysts and the zone of darkened porphyry grading into normal grey or yellowish

porphyry with abundant phenocrysts of quartz and feldspar. Basic dykes and sheets cut the porphyry and the whole intrusion is traversed by a NW–SE trending dyke to the south of the headland.

At low-water passage along the shore is possible. At high-water the cliff top should be gained either by a short scramble over porphyry or by ascending by way of a path of sorts up a little spur a short distance north of the conspicuous high face of porphyry. In either case a little care is necessary. The shore is readily regained by descending along an obvious gully eroded in the eastern marginal tholeiite. Should this excursion be made from south to north the descent from the cliff top, if necessary, should be made with great care.

South of An Cumhann the cliff has been hollowed out into caves, cut during Postglacial Raised Beach times. The rocks are sandstones, mainly grey or yellowish; beds of red sandstone also occur.

The largest of the caves, known as King's Cave, is now in the charge of the Department of the Environment and is closed by a railing. The cave gets its name from the legend that it sheltered King Robert the Bruce when he was hiding in Arran. The name of Fingal has also been associated with it. Apart from the carvings on the wall it seems to have yielded little of archaeological or antiquarian interest.

4. A short distance south of the caves there is a recess in the cliff the north side of which shows sandstones surmounted by a thick sill of felsite which is mostly deeply weathered to sandy material, red and yellow in colour. This sill is covered by sediments which are topped by a sill of pitchstone. Pitchstone is seen again at several points in the recess and, most readily, at the base of the conspicuous knoll on its south side. Somewhat devitrified pitchstone forms a low cliff below the footpath here. For some distance to the south a felsite-pitchstone sill forms a feature at or near the top of the cliff.

The junction between the Lamlash Beds and the Auchenhew Beds is almost certainly a fault running NW–SE along the line of a little stream in the recess as noted by Tomkeieff. It can readily be appreciated that the sediments to the north are predominantly sandstones while to the south of the recess siltstones and other fine-grained sediments, generally referred to in the literature on Arran as marls, figure largely.

Between the recess and Cleiteadh nan Sgarbh the shore shows good exposures of the Auchenhew Beds. Here, these are red and green marls with thin sandstones. A few small faults can be traced and three basic dykes cross the shore, two having considerable hades to the northwest, and a third, the most southerly, showing good exfoliation. Fallen blocks at the foot of the cliffs show a range of dessication features and trace fossils.

5. The promontary of Cleiteadh nan Sgarbh is formed of thick dykes of quartz-feldspar-porphyry and felsite associated with marginal tholeiitic dolerites. The acid rocks are 15·5 m and 9 m thick respectively; both intrusions appear to hade to the east. As they approach H.W.M., they swing round to the southeast, cross the raised beach and are last seen cutting the cliff. Near H.W.M. they are cut by an E–W dyke of dolerite. To the north of this dyke the porphyry is cut by an inclined sheet of dolerite which does not cut the felsite. The latter rock shows flow-banding; on weathering it takes on bright-red and yellow colours and locally looks deceptively like a sandstone. The porphyry is, as usual, darkened and xenolithic against the marginal dolerites.

6. The track which leads up from the raised beach should be taken to the north end of the Drumadoon cliff, thereafter the path along the talus slope under the cliff should be followed. From it the marly sediments underlying the sill, and also the basal parts of the Doon intrusion can be examined. The lowest member of the sill is a thin sheet of tholeiite above which rises columnar quartz-feldspar-porphyry in a sheer cliff some 25 to 30 m high. The lower levels of the porphyry are darkened and rich in xenoliths of dolerite which tends to weather out. This xenolithic porphyry is best examined in the fallen blocks and columns which litter the talus slope and the shore. If the summit of the Doon is visited the thin upper sheet of tholeiite, reported by Tyrrell (1928, p. 200) as being identified by fragments, may be looked for.

7. The shore west of the south end of the Doon cliff shows an east-west vertical margin of the composite intrusion. As usual marginal tholeiite is present, against darkened porphyry on one side and baked sediments on the other. Just to the north of the main contact a thin porphyry-tholeiite dyke runs off into the sediments on the shore.

The main porphyry makes an extensive spread on the foreshore southwards to Drumadoon Point. It is cut by several basic dykes against some of which the acid rock is in a better state of preservation than usual.

At Drumadoon Point the eastern margin of the intrusion can be examined. Here the tholeiite margin hades eastwards and contributes, in places, to an igneous breccia having a matrix of yellowish porphyry.

8. At this locality a dyke of quartz-feldspar-porphyry runs out to sea in a SSW direction. Its eastern margin is dark. The continuation of this dyke to the north, again with a dark eastern margin, makes a conspicuous feature on the golf course. This intrusion has been held to continue northwards along the east side of the Doon mass, perhaps acting as a feeder to it, and to connect with the porphyry of Cleiteadh nan Sgarbh, possibly even reappearing at An Cumhann.

In passing it should be noted that the golf course is laid out on blown sand as is often the case in coastal courses. These sands terminate inland against a raised beach cliff of marly sediments and intrusive rocks.

9. The shore south of the Club House shows marls and sandstone cut by numerous basic dykes. Traced to the east the dip of these sediments increases until it reaches 50° against the near-vertical margin of the Blackwaterfoot-Torr Righ Mor felsite. At the contact the felsite shows fluxion structures; a little to the east jointing gives it a roughly stratified and, in places, a "rippled", appearance. Elsewhere columnar joints, well seen in the Black Water at the bridge, are developed.

Dykes which trend in the direction of the felsite have irregular courses "as though they were influenced by the proximity of the felsite or found it difficult to penetrate it" (Tyrrell 1928, p. 222 and fig. 32).

Just south of the Club House the remains of two old sea-stacks stand in front of the raised beach cliff. Farther to the southeast the cliff shows a patch of sediments with columnar felsite to the west, and on its east side felsite again, but now with joints disposed in an asymmetric arch. This feature was noted and figured by Bryce (1872). The stacks and arch are situated in private grounds.

Excursion 14: **Area around Ard Bheinn**

(*Figure 18, localities 1 to 40*)

The purpose of this excursion is to examine the northwestern part of the Central Ring Complex, mainly around Ard Bheinn and Binnein na h-Uaimh. This area offers a great diversity of rock types and is for the most part fairly well exposed. The route map and the locality descriptions have been contributed by Professor B. C. King who has made a special study of the region (see King, 1954, 1955 and 1959). He regards the masses of sedimentary rocks and basic lavas within the Complex as relics of a block which subsided some 1,000 m during the formation of a volcanic caldera. The structure of the block has been complicated by dislocation along a number of internal arcuate and radial faults, as well as by extensive dissection as a result of the emplacement of intrusive agglomerates and breccias.

Further complexity has been introduced by the development of a series of lavas, breccias and agglomerates of andesitic and dacitic composition, together with extensive masses of felsite. These are regarded by Professor King as representing the basal parts of the superstructure of volcanic cones which rose above the caldera floor. The two main volcanic centres corresponded approximately to the positions of the Ard Bheinn and Binnein na h-Uaimh summits, while from the concentric and radial structures shown by the felsites other centres are inferred to the east and northeast.

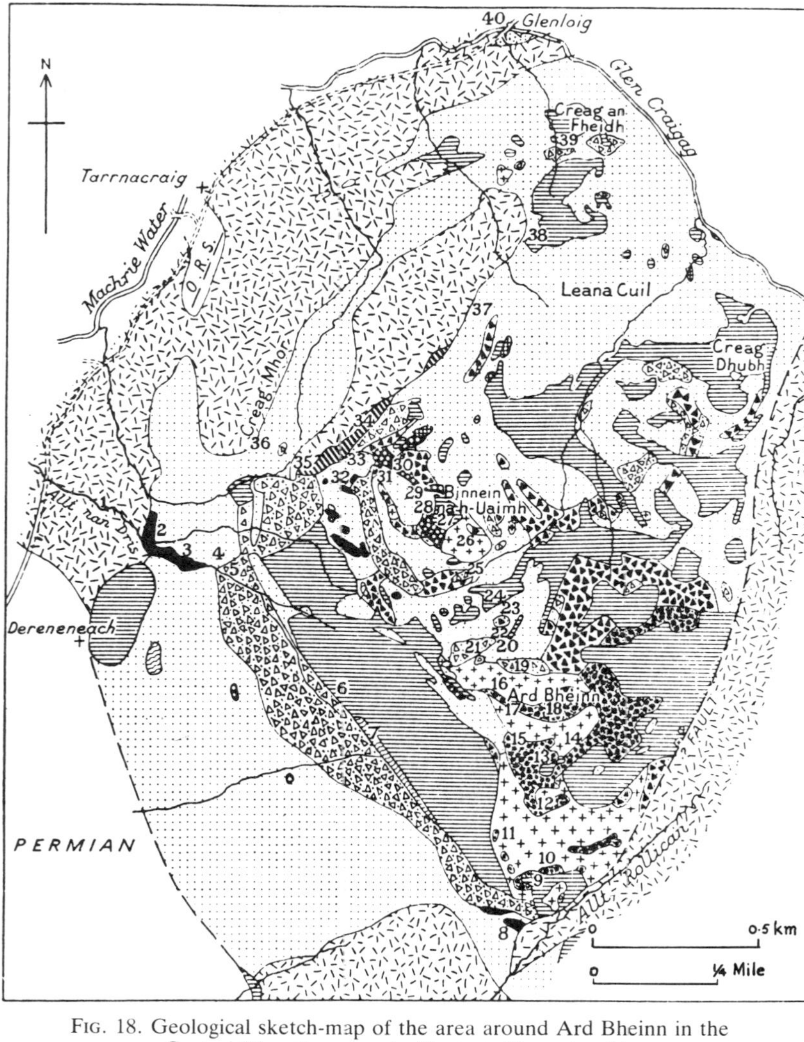

FIG. 18. Geological sketch-map of the area around Ard Bheinn in the Central Ring Complex, to illustrate Excursion 14.

A study of the two centres referred to is one of the main objects of the excursion. The lithology and the inter-relationships of the various rock formations are, however, extremely complicated and reference should be made to the detailed account of the area given by King (1955). The contours and crag featuring on the map accompanying this detailed account will help greatly in accurately locating particular outcrops.

The excursion, which includes visits to some of the masses of Mesozoic strata incorporated in the vent agglomerates, has been divided into two parts—localities 1–30 and 31–40.

Excursion 14a: **Allt nan Dris, Ard Bheinn and Binnein na h-Uaimh**

The excursion begins at the bridge over the Allt nan Dris about a quarter of a mile south of the junction of the Dougrie and Shiskine roads.

1. At this locality, some 230 m upstream from the bridge a small quarry provides a convenient exposure of a typical medium-grained granite (Tyrrell 1928, p. 192; King 1955, p. 345).

2 and 3. One hundred and fifty metres or so farther upstream, and along the contour at the sharp change of slope, occurs a series

---

INDEX TO FIGURE 18

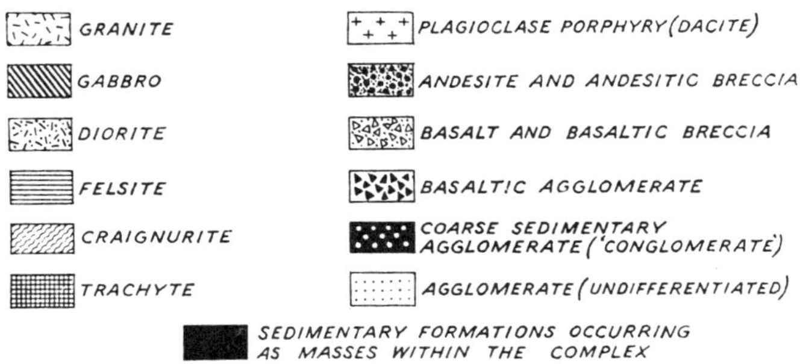

of small outcrops of characteristic red and grey Keuper marls. These pass upwards in a small gorge in the main stream at locality 3 into black shales with dark calcareous bands from which fossils of Rhaetic age have been obtained (Gunn 1903, pp. 73–75; Tyrrell 1928, pp. 105–107). The beds in question are considerably disturbed, but fragments of the diagnostic forms *Pteria contorta* and *Pecten valoniensis* have been found on recent excursions.

4. Examine here outcrops and boulders of the vent agglomerate. These represent the common type of agglomerate in the peripheral zone of the complex, characterised by an abundance of fragments and pebbles of sedimentary and metamorphic rocks set in a pale-weathering siliceous matrix. The agglomerate is formed almost exclusively of pre-existing formations, notably Dalradian and Old Red Sandstone.

5. Immediately south of locality 4 is the northern extremity of the belt of basalt lavas which extends southeastwards for about 1·5 km parallel to the margin of the complex. Steep craggy slopes define their northern and western limits and the straightness of the margins suggest that they may be tectonic. The basalts undoubtedly represent extrusions (not intrusions) and are closely comparable with the lavas of other Tertiary areas. Some shattering and shearing is often evident and in thin-section the rocks commonly show considerable alteration.

6. The basalts can be easily followed southeastwards as a steadily rising terrace on which the pattern of outcrops strongly suggests trap featuring. Note how they are bounded to the east by a sharply rising slope with numerous outcrops of pale-coloured felsite, contrasting strongly with the greenish-black and rusty-weathering basalts. The felsites (6) are flinty-textured rocks with small phenocrysts of quartz and felspar; weathered surfaces often suggest that they are brecciated.

7. Here a narrow band of craignurite is interposed between basalt and felsite and forms a marked topographic step. It is a distinctive rock, pale weathering but on fresh surfaces medium grey with small greenish mafic spots; it presents a fine-grained marginal phase towards the basalt.

8. This is the locality where Lias fossils were discovered by Gunn (1903, pp. 75–76; Tyrrell 1928, pp. 107–108). In his account he says: "The patch consists of brownish, crumbling, calcareous shale and impure decomposed limestone, in some parts of which fossils are abundant, but mostly in the form of casts. The exact area occupied by this shale cannot be ascertained, but it is probably 90 m in length and nearly 45 m in breadth." Characteristic fossils are easily obtained from the closely-jointed shales which outcrop in a steeply inclined gully.

9. Here conspicuous crags of a coarse-grained granite-porphyry crowded with phenocrysts of quartz and felspar form an obvious landmark. This type of rock is extensively developed on the eastern slopes of Ard Bheinn (King 1955, p. 337).

10. Examine at this locality outcrops of andesite and andesite breccia, easily recognisable by their dark-grey, almost black colour and paler-weathering crust. On the smooth, gentle slopes above, numerous platy fragments of plagioclase-porphyry can be seen. These relationships are typical of the rocks of the summit area of Ard Bheinn, the andesites forming dark, craggy outcrops, generally arcuate and following the contour, the porphyries poorly exposed and occupying the smoother slopes (King 1945, p. 341). Some difficulty may be experienced initially in distinguishing the two types, but the andesites show dark, rough-weathering surfaces and have a granular fracture, whereas the plagioclase-porphyry or dacite has a close platy jointing, a medium grey, sometimes black, colour and a flinty, occasionally glassy fracture. The andesites are seldom massive but their real fragmental nature may only be appreciated on weathered surfaces.

11. At this locality examine exposures of plagioclase-porphyry close to the junction with the main belt of felsite. The two rock types are easily distinguished by their colours. Note especially the occurrence near their junction of a crag of andesite. The relations between this and the felsite merit careful examination, for a narrow zone of streaky tuff intervenes in which the two rock types are mechanically mixed in varying proportion, sometimes with plagioclase-porphyry in addition (King 1955, p. 339).

12. At this locality, about 180m east-northeast of locality 11, there is a prominent line of crags composed of andesites, andesitic breccia and andesite-dacite breccia. These show a layered arrangement dipping southeast at 10° to 15°, while the most westerly outcrops consist of a rough block-lava of "aa" type (King 1955, Plate XIII). These outcrops should be carefully examined since they illustrate the evidence on which the writer bases the conclusion that these formations were deposited sub-aerially and belong to the volcanic super-structure.

13. Here there occurs a small area of quartz-feldspar porphyry surrounded by breccias of very varied composition (andesitic and mixed andesite and dacite). The porphyry differs from the usual felsites in having conspicuous phenocrysts. It is often autobrecciated and its northern margin shows a fine example of a banded tuff contact with andesite.

14. At this locality a closely mappable and arcuate junction between plagioclase-porphyry and andesitic breccia is exposed. The former passes into a dark, glassy, marginal phase with steeply inclined flow-banding and there can be little doubt that it is intrusive into the breccia.

15. The same glassy marginal phase can be seen here as fragments against crags composed of andesite and andesitic breccia.

16. Between localities 15 and 16 outcrops of andesite and plagioclase porphyry are bounded on the west by agglomerate containing numerous fragments of basalt and felsite.

17. Summit cairn. A line of crags extending from locality 16 to the summit shows typical andesite, ash-grey and pale weathering, with numerous amygdales of fine-grained quartz (chalcedony) often streaked by flow. Immediately to the north of the line of crags a junction made by dark glassy plagioclase-porphyry with steeply inclined flow banding is very well seen.

18. Good examples of typical andesites and andesitic breccia may be examined on the knolls east of the summit. From the summit a good view over the complex can be obtained; northwards lies the Binnein na h-Uaimh centre, while to the northwest the

numerous light-weathering crags indicate the abundant development of felsites.

19. The traverse continues down the gentle slope north of the summit for about 180 m across small exposures and boulders of plagioclase-porphyry succeeded at locality 19 by outcrops of basaltic breccia. These become andesitic westwards and pass into a diversity of andesitic (ash-grey in fracture and pale-weathering) and basaltic (greenish-black in fracture and rusty-weathering) breccias and agglomerates which occupy much of the northeastern slopes of the hill.

20. At this locality, between the two small outcrops of basaltic rock at localities 19 and 21 (q.v.) occur extensive exposures of agglomerate with conspicuous and often well rounded fragments of felsite.

21. Massive basalts and basaltic breccias are exposed beside the little loch at this locality.

The rock-types represented at localities 19 and 21 are interpreted by the writer as remnants of the updomed floor on which the Ard Bheinn volcano was built.

22. Here a dyke-like mass of felsite forms a prominent crag. Along its northwestern margin it shows against the agglomerate a transitional zone of felsite, breccia and tuff in all stages of mechanical mixture with the agglomerate (King 1955, p. 338). The details of the process can of course only be appreciated under the microscope.

23 and 24. The traverse down to the hollow just before the ascent to Binnein na h-Uaimh shows agglomerates commonly containing abundant fragments of felsite, often with basalt in addition, and sometimes with sedimentary debris also. At locality 23 a gently dipping layering can be seen, while below felsite forms an elaborate pattern of outcrops, the main mass of which, with highly tortuous boundaries can be traced for about 800 m to the east.

25. Here a ridge just beyond the hollow is seen to consist of basaltic breccias and agglomerates, the fragments including a wide

variety of basaltic types—olivine-basalt, tholeiite, and felspathic basalt. Andesites are also represented.

26. An extensive area of plagioclase-porphyry marks the approach to the summit. It is notable for the fact that locally it is made up of large blocks, often some metres across, although in places the blocky character is recognisable only from the haphazard orientation of the flow-banding.

27 and 28. The summit area is occupied by a felsite breccia and a remarkable agglomerate composed of large cobbles of felsite (27), while the steep western slopes show extensive outcrops of a coarse sedimentary agglomerate, resembling a conglomerate; indeed it is very largely made up of pebbles derived directly from Old Red Sandstone conglomerate (28).

29 and 30. The lithology and structure of the northwestern spur (29) and lower slopes (30) of the Binnein summit are exceedingly complicated and require lengthy examination. Basaltic breccia and agglomerate, sometimes containing very large blocks of basalt; actual masses of basalt; andesitic breccia; coarse sedimentary "conglomerate"; felsite agglomerate; and all possible variations of and gradations between, these types, show a crude layering inclined at about 20 degrees to northeast—an arrangement best appreciated at a little distance. Rapid lateral transitions occur between different types of agglomerate, but in places perfectly sharp, steep junctions are seen as at locality 30. Intrusive relations are again suggested by the sheeted mass of plagioclase-porphyry at locality 29. The impression conveyed is that of a diversity of deposits, partly sub-aerial, partly intrusive, varying from strictly volcanic products in origin to the deposits of volcanic landslips, the whole representing the elements of a volcanic superstructure, cut across during its development by intrusive porphyries and agglomerates.

Excursion 14b: **Ard Bheinn Area (Continued)**

31. Here, at the "Pigeon Cave", white crystalline limestone of Cretaceous age is seen passing under basalt, the latter forming the

northern extremity of a belt which can be traced for about 400 m along the western slopes of Binnein na h-Uaimh. The basalt is normally margined by agglomerate but at or near the lower boundary masses of a fine-grained sandstone occur at localities 32 and 33. Immediately below the latter gabbro is encountered; this is a rather fine-grained rock with a close jointing evidently related to the parallelism of the plagioclase crystals.

34. Along the southwestern margin of the gabbro is a finer-grained contact phase which proves on sectioning to be a tuff composed of fragmented gabbro (King 1954).

35. Here a contact with granite should be examined; the granite is obviously later than the gabbro since it veins and acidifies it.

36. After traversing the marshy depression to the west the margin of the Creag Mhòr granite is reached. Only at this locality can the contact margin be examined. The smoothly arcuate boundary elsewhere suggests that the granite intrudes the agglomerate, but here, careful search revealed a tiny patch in which agglomerate penetrates and fragments the granite. It is easily demonstrable also that the agglomerate contains many fragments of granite in the vicinity of the contact.

37. To reach this locality the marshy divide should be recrossed and the slope marking the eastern margin of the other granite mass followed. Here, in a series of prominent outcrops, the contact relations between granite and vent-agglomerate are well exposed. The agglomerates are strongly hornfelsed and re-crystallisation of the various fragments is clearly seen; ultimately the fragments become shadowy and indistinct and the matrix resembles a fine-grained granite or granophyre (King 1955, p. 346).

From locality 37 a detour may usefully be made to examine the felsites and their contacts in the area of Leana Cuil and Creag Dubh.

38. Examine the felsite at this locality; although it is later than the agglomerates it is demonstrable from the examination of thin sections that it was re-crystallised together with the adjacent agglomerates in the granite aureole.

39. Traverse the felsite until a small stream is reached which

makes a rapid descent down a steep gully. Here a second cave, enlarged by quarrying, has been formed in a mass of Cretaceous limestone, again overlain by basalt.

The great diversity of rock types in the vicinity of Creag na Fheidh suggests that it represents a minor volcanic centre.

40. Descend to Glenloig and just below the bridge examine good exposures of dioritic hybrid rocks, showing rapid variations in texture and composition.

5 ROAD LOGS

For the benefit and guidance of those who may wish, or even prefer, to become acquainted with Arran's geology by visiting and examining such rocks as are to be seen by the roadside, or within easy walking distance therefrom, the following "road logs" are offered. Drivers who do not know the island must be warned that parking, even for short periods, is often difficult or even not really feasible without creating dangerous conditions. Where appropriate, references to localities mentioned in Part 4 (the Descriptive Itineraries) are given. Parking space is usually available at the numbered localities given (see figure 1).

Road Log No. 1: **Brodick — The String — Blackwaterfoot — Pirnmill — Lochranza — Corrie — Brodick (A841 — B880 — A841)**

1. The shore west of the ferry terminal provides good exposures of the Brodick Beds, these being cut by several dolerite dykes showing the two small areas of brecciation described by Tyrrell as "cryptovolcanic" (Excursion 1, loc. 1). The larger of the latter lies near H.W.M. close to a roadside shelter.

2. With permission the pitchstone at the old school (010366) may be examined. Please do not hammer here (Excursion 1, loc. 2). Proceed to the A841–B880 junction and follow the later road to the west towards Blackwaterfoot.

3. On the south side of the road at (983360) is an old quarry in steeply dipping sandstone, much shattered and cut by a NE–SW dyke. The sedimentary rocks here belong to the Lower Old Red Sandstone.

Overlooking the summit of the road on its south side is a prominent crag formed by crushed, sheared and partly amphibolitised gabbro which locally is seen to rest on baked Old Red Sandstone and to be cut by veins and dykes of granophyre and aplite.

4. At the bridge over the Allt nan Calaman (969354) fine-grained doleritic rocks are seen, both in the stream and in a roadside excavation. The little black prisms in these rocks are mainly hornblende, the white matrix being largely composed of feldspar.

5. At Glenloig farm (946351) the stream which issues from Glen Craigag exposes fine-grained diorites (Excursion 14b, loc. 40). The hillside to the south provides exposures of several of the rock types found in the Central Ring Complex (Excursion 14b, locs. 38–40). The imposing crags which overlook Glen Craigag from the east are formed by volcanic agglomerate.

From the end of the road to Machrie, at a remarkable old pillar box (946351), the fine scar of Craig Mhor is well seen to the east. The rock is the outer granite of the complex.

6. The above granite can be examined on the banks of the Allt nan Dris (931336; Excursion 14a, loc. 1).

Southwards from Ballymichael the road runs just to the east of the largest area of alluvial deposits on the island. These are the marine alluvia of the High Lateglacial Raised Beach (the 100 Foot Raised Beach of the I.G.S. maps) covered in part by more recent fresh water alluvium.

7. At Blackwaterfoot the felsite, and other rocks mentioned in Excursion 13 (loc. 9) can be examined conveniently close to the point at which the road turns north. *En route* to Machrie the granite hills of the northern part of the island come into view with Beinn Bharrain (721 m) in the west followed eastwards by the isolated and lower summit of Sail Chalmadale (481 m).

8. On the shore and to the north of the road junction at Machrie steeply dipping sandstones and conglomerates are cut by dykes and sill-like intrusions of dolerite, felsite, porphyry and pitchstone, some of which are exposed only at low water. The steep dip of the sediments is particularly well seen in the raised beach cliff.

The New Red Sandstone can be distinguished from the Upper Old Red Sandstone by the presence of small pieces of agate.

9. Northwards, towards Dougrie Lodge, conglomerate and sandstone of the Lower Old Red Sandstone are well seen both on the shore and in the raised beach cliff.

The shore south of Ballykine farm (870397) provides exposures of fine-grained dark schists which occupy the core of the Loch Ranza antiform (p. 23). To the south of this locality the more gritty and, locally, pebbly schists can be proved from graded bedding to have normal dips, while to the north of it the beds are inverted.

Note: A graded bed is a sedimentary unit relatively coarse in grain size at its base and becoming progressively finer towards its top. Such a unit, or better, a succession of such units is of great value in determining the stratigraphical order in folded metamorphic rocks.

10. Just south of Imachar Point a large dyke of quartz-dolerite (of late Carboniferous age) cuts the Dalradian schists on the shore and in the raised beach. This intrusion is the most southerly known representative in Scotland of thick, generally east-west trending, tholeiitic dykes which extend in a zone northwards as far as Aberdeenshire. Similar dykes are associated with the Whin Sill of the north of England.

Rather less than 1 km north of this dyke and at the remains of a fence (862412) the schists are cut by a narrow NW–SE trending composite dyke on the NE side of which the schists have been melted locally. The blackish glass which formed when the rocks cooled can be traced for up to 20 cm from the contact (Holgate 1978, 172–4). **No hammering here, please!**

11. At Pirnmill gritty schists, dipping rather steeply westwards, are readily examined. For details of the rock types see Excursion 9 (loc. 17) and Excursion 10 (locs. 19, 22 and 23). Although gritty schists (formed from coarse sandstones and fine conglomerates which have suffered mechanical deformation and some chemical reconstitution) predominate, the shore section includes beds of

finer grained material, probably originally shales, in which metamorphism is more marked. In these beds white mica gives the foliation planes a glossy appearance, while chlorite imparts a greenish colour to some of the rocks.

12. The Catacol Burn (910489), upstream from the main road, exhibits a fine section through the remains of what was probably a glacial outwash fan deposited by meltwater from the glacier which occupied Glen Catacol in Lateglacial times. This is followed by a little gorge in steeply dipping, highly baked schists cut by dykes. Rather less than 1 km from the road the granite/schist contact crosses the burn: although the actual contact is not exposed there it can be located closely (Excursion 9, locs. 3–5).

13. As the road is followed from Catacol to Lochranza it can be seen that the dip of the schists, here mainly gritty, changes from northerly to southeasterly as the axis of the fold, known as the Catacol Synform, is crossed (Excursion 10, loc. 29; fig. 13).

14. The car park at North Sannox Bridge (994468) is a suitable location from which to examine (*a*) upstream from the bridge, the Dalradian schists and the sedimentary features exhibited by them (Excursion 6, loc. 11); (*b*) approximately 1 km west of the bridge a well exposed granite/schist contact (Excursion 6, loc. 10); and (*c*) downstream from the bridge, the pillow lavas and associated black shales and cherts of the Ordovician succession (Excursion 6, locs. 14–16).

From Sannox southwards to Brodick Old Quay, the shore, and in places the adjacent raised beach cliffs, allow ready examination of, successively, the upper part of the old Red Sandstone (Excursion 6, loc. 22; Excursion 4, locs. 1–3); the Carboniferous succession in this part of the island (Excursion 4, locs. 4–12); the Corrie Sandstone facies of the Brodick beds intruded by dykes which are especially numerous opposite Corrie Terraces (026424) and southwards from Rudha (or Rubha) Salach to the Old Quay. Parking varies from easy to impossible.

It is suggested that the following parts of the Corrie shore section are likely to prove of most interest to the visitor who is short of time.

(*a*) The conglomerates, cornstones, volcanic agglomerate etc., at the Upper Old Red Sandstone/Lower Carboniferous junction (023441; Excursion 4, locs. 3 and 4).

(*b*) The olivine basalts, with phenocrysts of augite and olivine, north of the school. Weathering into spherical masses is shown in these rocks.

(*c*) The Corrie Limestone, seen in mines west of the road opposite the harbour. For details and a warning which should be heeded see Excursion 4 (loc. 8).

(*d*) The sandstones, thin limestones and associated strata a little north of the Corrie Hotel. These rocks are of Carboniferous Limestone age.

(*e*) In the little bay just south of the hotel, sandstones, fireclays and shales belonging to the Upper Carboniferous outcrop. Fossil plants and fresh-water "mussels" are locally present.

(*f*) The unconformable junction of the dune-bedded Corrie Sandstone (brick red) on Upper Carboniferous sandstone (whitish, with contorted bedding) occurs a little south of the Corrie Hotel (026431). A thin sheet of basalt cuts the Carboniferous a little below the unconformity.

Road Log No. 2: **Brodick — Lamlash — Whiting Bay — Kildonan — Bennecarrigan (A841) returning to Lamlash via the Ross Road (unclassified)**

En route to the summit 3 km south of the ferry terminal, red sandstones and boulder clay are commonly exposed on the east side of the road. Dykes can also be seen, e.g. just south of the Corrygills road (Excursion 3, loc. 1).

15. At the summit (018334) where a Forestry Commission road joins, the analcime-dolerite of the Clauchlands intrusion is exposed rather poorly (Excursion 3, loc. 11). From this point there

is a good view of the Northern Granite mountains (Plate 1), and to the southeast, Holy Island and Lamlash Bay. Holy Island is again well seen from the road to Whiting Bay. That it is formed of columnar igneous rock (trachyte) overlying red sandstone can readily be appreciated.

16. Between Kingscross Bridge (043280) and the road to Auchencairn (044278) the following section can be examined on the west side of the road. Just south of the bridge, and near road level, coarse analcime-dolerite is seen. This is followed southwards by drift beyond which is an interesting section comprising a composite felsite-quartz-dolerite sill overlying and locally in transgressive relationship to sandstones which are intruded by an altered dolerite. Another stretch showing drift only is followed to the south by an extensive exposure of analcime-olivine-dolerite. Just north of the Auchencairn road junction this intrusion includes a few small rafts of sandstone.

17. Especially at low water the shore at the north end of Whiting Bay (047270) shows a fine example of a plexus of intersecting dykes, crinanitic and tholeiitic types being represented. The sediments are soft red sandstones with ripple marks and other sedimentary structures.

18. **(with care)** Creag Dhubh (049250) is formed by a neat example of a composite sill. Just under 10m thick it, and the sandstone into which it has been intruded, dip at 20° to the WSW. The body is composed of a central layer of quartz-dolerite flanked by felsitic layers, with merging contacts.

19. A short distance south of the houses at Largybeg variegated marly sediments (which have yielded microfossils of Triassic age) are cut by a rather irregular sill of dark, fine-grained quartz-dolerite. Above lies the rudely columnar Dippin sheet of coarse analcime-olivine-dolerite. This rock is exposed in two small roadside quarries near the 12 mile stone. In both quarries the sheet is cut by dolerite dykes.

By taking the unclassified road to Kildonan part of the Arran dyke swarm is readily seen, especially at low tide. In this area the dykes run mainly NNW–SSE and they vary greatly in width.

20. A particularly massive dyke (of crinanite) occurs near the point (019210) at which the road turns inland (see Excursion 12, loc. 9). A short distance offshore lies Pladda island, mainly occupied by a sill of quartz-dolerite; the low cliff on this island is the result of marine erosion at the time of the Main Post glacial Shoreline when sea level was higher than at present (see pp. 63–66). In the distance, to the southeast, lies Ailsa Craig, an island formed entirely of a plug of alkali microgranite with petrographic affinities to the Holy Island trachyte.

En route past Auchenhew look out for Auchenhew Hill and Levencorroch Hill, lying to the north of the road. They are capped by a dolerite sill which has been cut through by a deeply incised stream, the Levencorroch Burn.

21. Close to the junction of the private road to Craigdhu farm with the main road (000213) is an old quarry in the central, acid member of the Bennan composite intrusion. This is a quartz-feldspar-porphyry composed of phenocrysts of feldspar (usually dull white) and quartz (dark, glassy) in a finer-grained pale-coloured matrix. Nearby, the western doleritic marginal member outcrops by the farm road (Excursion 12, loc. 12), dipping steeply westwards.

The main road is now followed via Shannochie and Lagg to the south end of the Ross Road at the church (943220) near Bennecarrigan farm.

Rather more than a mile north of the church in a quarry (947239) is a badly preserved felsite. In the vicinity of the bridge over the Sliddery Water south of Glenree farm (946252), or Burrican as it is named on older maps, the stream provides sections in two felsites, one of which is associated with pitchstone.

Almost on the summit of the road and on its northwest side is a small quarry in which an "igneous breccia" composed of angular fragments of basic rocks are included in a matrix of pale-weathering silicic rock. Tyrrell (1928, p. 140) has sketched a specimen from this locality. Please do not hammer here: loose material can be readily had from the adjacent hillside.

In the vicinity of the 8 mile stone (994300) there are good exposures of cross-bedded, sometimes pebbly, sandstone cut by dykes of dolerite and felsite. The old quarry just east of the milestone shows sandstone cut by four dykes one of which is a porphyritic dolerite and the other, felsite. The view looking eastwards from a little farther down the road shows that, on both sides of the glen, the base of the Monamore intrusion slopes evenly downwards towards the east. The most westerly crag on the north side of the road has a banded appearance due to the presence of rafts of baked sandstone in the igneous rock.

22. The picnic area within Forestry Commission ground a short distance west of Croc is close to the eastern margin of the Monamore analcime-olivine-dolerite intrusion. A traverse of the Monamore Burn shows highly-baked conglomerate followed upstream by the intrusion which is cut by several dolerite dykes and includes screens of sandstone. At the remains of an old footbridge there is a small body of quartz-felsite. Farther upstream a much larger body of granophyric felsite occupies the banks and bed of the burn. The rock of the main intrusion has smooth, rounded surfaces in which pot holes have been formed.

Intrusions of felsite and pitchstone are to be seen cutting sandstone in the Monamore Burn almost opposite the road to Glenkiln (Glenkill on older maps).

6 GLOSSARY OF PLACE-NAMES

The place-names of Arran are in the main derived from Gaelic sources. Some, however, are of Norse (more exactly Old Norse) provenance and these are indicated in the following list by the letter N. Sometimes both sources contribute to the name as in Corrygills; and sometimes the Gaelic *Gleann* (Glen) has been added to the Norse names of valleys, e.g. Glen Chalmadale, Glen Ormidale, Glen Scaftigill, etc. Some names, again, are translations of the original Gaelic, like Blackwaterfoot, Birch Burn, etc. In a number of cases the etymology of the place-names is doubtful and these have been mostly omitted.

The following notes on Pronunciation are taken from "Glossary of the most common Gaelic Words (and Corrupted Forms of Gaelic) used on the Ordnance Survey Maps", published by the Ordnance Survey.

Initial Bh or Mh equals V, but after a broad vowel equals W, as in English "now".
,, C equals K.
,, Fh is silent.
,, Ph equals F.
,, Sh or Th equals H.
,, S after An t is silent.
Final aidh equals y as in my.
,, idh equals y as in duty.
Th final, or when flanked with vowels, is a strong breathing.
Ch in contact with a, o, or u is a strong guttural as in *loch*.
Ch in contact with e or i is a guttural as in German *ich*.

The Article:
A' equals *the*, as in A'Chir.

Am equals *the,* as in Am Binnein
An ,, ,, as in An Cumhann.
A' ,, *of the,* as in Cnoc a'Chapuill.
Na ,, ,, as in Cnoc na Ceille.
Nan ,, ,, as in Torr nan Gobhar.
Nam ,, ,, as in Uisge nam Fear.
An t- ,, ,, as in Torr an t-Sean Chaisteal.

ABHAINN BEAG: *abhainn,* a river; *beag,* small.

A'CHIR: *cìr,* a crest or comb. This ridge is often referred to as "The Sleeping Warrior".

ACHAVOULIN: *achadh,* a field or piece of land; also in the forms *ach, auch; muileann,* a mill.

ACHNAMARA: *muir* (genitive *mara*), the sea.

A'CHRUACH: *cruach,* a stack or pile.

ALLT A'CHAPUILL: *allt,* a burn; *capull* (genitive *chapuill*), a horse.

ALLT BEITHE: *allt,* a stream or burn; *beithe,* a birch-tree.

ALLT CAIRN BHAIN: *càrn* (genitive *cairn*), a heap of stones; *bàn,* white.

ALLT DIOMHAN: perhaps from *deamhan,* a demon.

ALLT GARBH: *garbh,* rough.

ALLT GOBHLACH: *gobhal,* a fork; i.e. the forked stream.

ALLT MÒR (or MHÒR): *mòr,* large.

ALLT NAN DRIS: *dris,* bramble or brier bush.

AM BINNEIN: *binnein,* a pinnacle; from same source as *beinn.*

AN CUMHANN: *cumhann,* narrow; the narrow place.

AN SGRIOB: *sgrib,* a rent or cleft.

AN TUNNA: *tunna,* a cask.

ARD BHEINN: *àrd,* high.

AUCHAGALLON: *gallan,* a standing stone.

AUCHAREOCH: *riabhach,* dappled or bridled; see under *achadh.*

AUCHENCAIRN: the field of the cairn.

AUCHENCARR: *carr,* a rock.

AUCHENHEW: *eó,* holly.

AUCHMORE: *mòr,* large.

BALLARIE: *baile,* a dwelling; and perhaps *airidh,* a sheiling.

BALLYMICHAEL: Michael's dwelling.

BEALACH AN FHARAIDH: *bealach,* a mountain pass; *faradh,* a ladder.

BEALACH AN FHIR BHOGHA (the Archer's Pass): *fear* (genitive *fhir*), a man; *bogha,* a bow.

BEINN A' CHAISTEAL: *beinn,* a mountain; *caisteal,* a castle.

BEINN A' CHLIABHAIN: *cliabh,* a basket or cradle.

BEINN (Ben): a mountain.

BEINN BHARRAIN: *bearna,* a gap or cleft; Ben Bharrain has two peaks.

BEINN BHREAC: *breac,* dappled.

BEINN CAORACH: *caora* (genitive *caorach*), a sheep.

BEINN NUIS (*Beinn nan Ois*): *os,* a fawn.

BEINN TARSUINN: *tarsuinn,* transverse or across, i.e. the crossing or joining hill.

BENNAN: a diminutive of *beinn.*

BINNEIN NA H-UAIMH: *Binnein,* a pinnacle; *uamh,* a cave.

BIRCHBURN: see Allt Beithe.

BLACKWATERFOOT (in Gaelic *bun na abhainn dubh*): *bun,* the end or mouth; *abhainn,* a river; *dubh,* black.

BLAIRBEG: *blàr,* a mossy tract; *beag,* small.

BLAIRMORE: *mòr,* large.

BRODICK: N. *breid-r,* broad; *vik,* a bay.

CAISTEAL ABHAIL: *caisteal,* a castle; *gobhal* (genitive *ghabhail*), a fork.

CATACOL: N. *kött-r,* a cat; *geil* or *gil,* a ravine; i.e. ravine of the wild cat.

CEUM NA CAILLICH: *ceum,* a step; *cailleach* (genitive *cailliche*), an old woman.

CIOCH NA H'OIGHE: *cioch,* a pap; *oigh,* a maiden.

CIR MHOR: a crest or comb; *mòr,* great.

CLACH A'CHAIT, CLACH AN FHIONN and CLACH MHÒR are large granite boulders on the Corrie shore; *cat,* a cat; *fhionn,* Fingal, and *mòr,* large.

CLACHAG (the stony place): *clach,* a stone.

CLACHAN (the stone church): from *clach,* a stone; St. Molios is said to have preached here.

CLAUCHAN (Clachan) GLEN and CLAUCHLAND (Clachland) are also from the same word.

CLEITEADH NAN SGARBH: *cleiteadh,* a rocky promontory; *sgarbh,* a cormorant.

CNOC (pronounced krock) A'CHAPUILL: *cnoc,* a hillock; *capull* (genitive *chapuill*), a horse.

CNOC LEACANN BUIDHE: *leacann,* a bare hillside; *buidhe,* yellow.

CNOC NA CEILLE: *ceilidh,* a place of meeting.

CNOC NA CROISE: *crois,* a cross.

CNOC NA GARBAD: *garbh,* rough; *bad,* a thicket.

CNOCAN DONN: *cnocan,* a diminutive of *cnoc; donn,* brown.

CNOCAN DONNA: *donnaidh,* a misfortune.

COILLEMORE: *coille,* a wood; *mòr,* large.

COIRE AN LOCHAN, the corrie of the little loch.

COIRE NAN LARACH: perhaps from *làir* (plural *làrach*), a mare.

CORLOCH: *còrr,* a heron or crane.

CORRYCRAVIE: *coire,* a hollow or cauldron; *craobh,* a tree.

CORRYGILLS: Gaelic *coire* and Norse *geil* or *gil,* a ravine; the "*s*" is a later addition.

CREAG BHAN: *creag,* a rock; *bàn,* white.

CREAG DUBH: *dubh,* black.

CREAG NA FEIDH: *creag,* a rock; *fiadh,* a deer.

CREAG NA FITHEACH: *fitheach,* a raven.

CREAG NA H-IOLAIRE: *iolair,* an eagle.

DERENENEACH: *doire,* a thicket; *each,* a horse.

DIPPIN: *dubh,* black; *peighinn,* a penny (land); derived from the old system of measuring land by rentals; there is a Pennyland in Kintyre.

DRUMADOON: *druim,* a ridge; *dun,* a hill-top fort; i.e. the stronghold on the ridge.

DUBH LOCH: *dubh,* black.

DUN DUBH: *dun,* a hill-top; *dubh,* black.

DUN FIONN, Fingal's Fort.
DUNAN: a diminutive of *dun*.

EAS A'CHRANNAIG: *eas,* a waterfall; *crann,* a tree; the ending *aig* implies abundant.
EASAN BHIORACH: see under GLEN.

GARBAD: *garbh,* rough; *bad,* a thicket.
GARBH ALLT: *garbh,* rough; *allt,* a stream or burn.
GARBH THORR: *tòrr,* a small round hill.
GLEN or GLEANN ASHDALE: *eas,* a waterfall; *dal-r,* a valley; the Gaelic descriptive term *gleann* (glen) has been added.
GLEN CHALMADALE: here again the Gaelic *gleann* has been added to a place-name of Norse origin.
GLEN CRAIGAG (glen of the little rock): *creag,* a rock.
GLEN DUBH: *dubh,* black.
GLEN EASAN BHIORACH: *eas,* a waterfall; and perhaps *biorach,* a heifer.
GLEN ORMIDALE: N. *orm-r,* a snake; *dal-r,* a valley.
GLEN ROSA: the ending "*a*" is the N. *à,* a river; the first part of the name is given by different authorities as N. *hross,* a horse, or Gaelic *rosach,* reddish or rosy.
GLEN SANNOX: see under SANNOX.
GLEN SCAFTIGILL: N. *skaft,* a shaft; *à,* a river; *geil* or *gil,* a ravine; here again the Gaelic *gleann* has been added.
GLEN SHANT: the charmed or blessed glen.
GLEN SHURIG: probably *searrach,* a foal.
GLENLOIG: *laogh,* a calf.
GOATFELL: probably N. *geit,* a goat; *fjall,* a mountain or "fell".

HOLY ISLAND: the old name was *Eilean Molais* (St. Molio's isle), of which Lamlash, the name now applied to the village is said to be a corruption.

IMACHAR: *iomaire,* a ridge or division of land, with reference to the old run-rig system of cultivation.

INVERCLOY: *inbhear* or *inbhir,* the mouth of a river.
IORSA: N. *iór,* a horse; *à,* river.

KILBRANNAN SOUND: *caol,* a strait or sea passage; i.e. the Strait of St. Brendan (see p. 74). To the Gaelic name the Norsemen added their own term *sund.*

KILBRIDE, the church of St. Bridget: *ceall* (dative *cill*), a hermit's cell, or place of worship.

KILDONAN, KILMICHAEL, KILMORY: see above.

KING'S CAVE: the Gaelic name is *Uamh an Righ,* cave of the king.

KINGSCROSS: this name and the preceding one are associated with Bruce.

LAG or LAGG: *lag*, a hollow.

LAG A'BHEITHE: *beithe,* a birch tree.

LAGGAN: a diminutive of *lag.*

LAMLASH: see under HOLY ISLAND.

LARGYBEG, LARGYMEANOCH, LARGYMORE: *learg,* a hillside; *beag,* little; *meadhonach,* middle; *mòr,* large.

LENNYMORE: *leanach,* a wet meadow or piece of land; *mòr,* large.

LEVENCORROCH: the rough halfpenny land; *leth,* half; *peighinn,* a penny (land); *corrach,* rugged.

LOCH A' MHUILLIN: *muileann,* a mill.

LOCH COIRE AN LOCHAN: see under COIRE.

LOCH GARBAD: *garbh,* rough; *bad,* thicket.

LOCH NA DAVIE: *da,* two; *abh,* (*ainn*), a river; i.e. loch of the two rivers.

LOCH TANNA: *tana,* shallow.

LOCH URIE (LOCH H-IUBHRAIDH): *iubhar,* the yew tree.

MACHRIE: *machair,* a flat plain, carse land.

MADADH LOUNIE: *madadh,* dog or wolf; lounie from (*al*)*luidh-n-aigh,* or *allaidh,* fierce or wild; *aigh,* place of.

MAOL DONN: *maol,* a bare round hill; *donn,* brown.

MEALL BHREAC: *breac,* speckled.

MEALL NAN DAIMH: *meall,* a shapeless hill; *damh,* a stag.

MERKLAND: the merk was an old silver coin; the sum paid for rental.
MONAMORE: *moine,* a moor; *mòr,* large.
MONYQUIL: probably the hazel moorland; *coll,* the hazel.
MUILEANN GAOITHE (Windmill Hill): *muileann,* a mill; *gaoth,* the wind.
MULLACH BEAG and MULLACH MÒR (Holy Island): *mullach,* the top.

NARATHAN (the place of snakes): *nathair,* a snake.

PENRIOCH: *peighinn,* a penny (land); *riabhach,* brindled.
PLADDA: N. *flatr,* flat; *ey,* an island.
PORT A' GHILLE: *port,* a harbour; *gille,* a lad.
PORT A' LEACACH: *leac,* a flat stone; i.e. harbour of the many flagstones.
PORT NA GAILLIN: *port,* a harbour; *gaillionn,* a storm.

RUDHA BHAN: *rudha,* a promontory; *bàn,* white.
RUDHA CREAGAN DUBH: promontory of the little black rock.
RUDHA GLAS: *glas,* grey.
RUDHA SALACH: *salach,* dirty.

SAIL AN IME: *sail,* a heel; often applied to the end of a high ridge; *ìm,* butter.
SANNOX (the sandy bay): N. *sand-r,* sand; *vik,* a bay; the "*x*" is later addition; pronounced sannaig in Gaelic.
SCRIDEN: *sgriòdan,* a landslide.
SHEEANS: *sith,* a fairy; *sithean* is the fairy home.
SHISKINE: *sescem* (pronounced shesken), a marsh.
STACACH: *stac,* a steep conical hill.
STRABANE: *strath,* a strath; *bàn,* white.
STRATHWHILLAN: *cuilionn,* holly.
STRING ROAD: N. *strengr,* a cord or bridle; the String must have been merely a rough track.
STRONACH: *sròn,* a nose.

STRUEY: *sruth,* a stream; termination *ey* for *aidh* means the place of.

SUIDHE FHEARGHAS (the resting place of Fergus): *suidhe,* a seat.

TIGHVEIN: *taobh,* a hillside; *beinn,* great.

TORBEG: *beag,* small.

TORMORE: *mòr,* large

TORR AN T-SEAN CHAISTEAL: *tòrr,* a round hill; *sean,* old; *caisteal,* a fort.

TORR BREAC: *breac,* dappled.

TORR DUBH: *dubh,* black.

TORR NA BAOILEIG: *baoilaig,* blaeberry.

TORR NAN GOBHAR: *gobhar,* a goat.

TORR REAMHAR: *reamhar,* fat.

TRAREOCH (Traghariabhach): *tràigh,* the shore; *riabhach,* speckled.

UISGE NAM FEAR: *uisage,* water; *fear,* a man.

7 REFERENCES

The *Geological Survey Memoir of 1928* (pp. 277–283) contains an exhaustive bibliography of publications dealing with Arran. Any papers on the geology of the island that have appeared since that date are included in the following list of publications to which reference is made in the text of the Excursion Guide.

Allport, A. 1872. On the microscopic structure of the pitchstones of Arran. *Geol.Mag.*, **9,** 1–10.

Allport, A. 1872a. On the microscopic structure of the pitchstones and felsites of Arran. *Ibid.*, **9,** 536–545.

Anderson, J. G. C. 1945. The Dalradian rocks of Arran. *Trans.geol.Soc.Glasgow.*, **20,** 264–86.

Anderson, J. G. C. 1947. The geology of the Highland Border; Stonehaven to Arran. *Trans.roy.Soc.Edinb.*, **61,** 479–515.

Anderson, J. G. C. 1954. Hutton's Unconformity. *Geol.Mag.*, **91,** 85–86.

Anderson, J. G. C. and J. Pringle. 1944. The Arenig rocks of Arran and their relationship to the Dalradian series. *Geol. Mag.*, **81,** 81–87.

Ashford, J. H. 1959. The Structure and Petrology of the Eastern Part of the Central Igneous Complex of Arran. *Unpubl. Ph.D.Thesis, Univ. of Glasgow.*

Astin, T. R. and D. I. M. Macdonald. 1983. Syn-depositional and valley-fill breccias in the Permo-Trias of Arran. *Scott.J.Geol.*, **19,** 47–58.

Bailey, E. B. 1926. Subterranean penetration by a desert climate. *Geol.Mag.*, **63,** 276–280.

Bailey, E. B. 1926a. Domes in Scotland and South Africa: Arran and Vredefort. *Ibid.*, **63,** 481–495.

Balfour, J. A. and others. 1910. *The Book of Arran.* Glasgow.

Barrett, B. H. 1925. The Permain Breccia of Arran. *Trans.geol. Soc.Glasg.*, **17,** 264–270.

Bluck, B. J., A. N. Halliday, M. Aftalion and R. M. Macintyre. 1980. The age and origin of the Ballantrae ophiolite and its significance to the Caledonian orogeny and Ordovician time scale. *Geology,* **8,** 492–495.

Bonney, T. G. 1877. On certain rock-structures, as illustrated by pitchstones and felsites in Arran. *Geol.Mag.,* **14,** 499–511.

Bremner, A. 1942. The origin of the Scottish river system. *Scott.geogr.Mag.,* **58,** 15–20, 54–59, 99–103.

Briggs, D. E. G., W. D. I. Rolfe and J. Brannan. 1979. A giant myriapod trail from the Namurian of Arran, Scotland. *Palaeont,* **22,** 273–291.

Bryce, J. 1872. *The Geology of Arran and the other Clyde Islands (4th Edition).* Glasgow and London.

Bryce, T. H. 1910. The sepulchral remains of Arran. *In* Balfour and others, 33–155.

Cadell, H. M. 1886. The Dumbartonshire Highlands. *Scott.geogr. Mag.,* **2,** 337–347.

Cadell, H. M. 1913. *The Story of the Forth.* Edinburgh.

Callander, J. G. 1929. Land movements in Scotland in prehistoric and recent times. *Proc.Soc.Antiq.Scotland,* **63,** 314–322.

Challinor, J. 1961. *Dictionary of Geological terms.* Cardiff.

Cowie, C. R. 1905. Notes on the Shoals at Catacol Point and Kirn Point, Arran. *Trans.geol.Soc.Glasgow,* **12,** 156–158.

Craig, G. Y. 1965. *The geology of Scotland.* Oliver and Boyd, Edinburgh.

Curry, G. B., J. K. Ingham, B. J. Bluck and Alwyn Williams. 1982. The significance of a reliable Ordovician age for some Highland Border rocks in Central Scotland. *J.geol.Soc.London,* **139,** 451–454.

Dagley, P., A. E. Mussett, R. L. Wilson and J. M. Hall. 1978. The British Tertiary igneous province: palaeomagnetism of the Arran dykes. *Geophys.J.R.astr.Soc.,* **54,** 75–91.

De Souza, H. A. F. 1982. Age data from Scotland and the Carboniferous time scale. In *Numerical Dating in Stratigraphy* (edit. G. S. Odin), Wiley.

Dewey, J. F. 1961. A note concerning the age of the metamorphism of the Dalradian rocks of Western Ireland. *Geol.Mag.,* **98,** 339–405.

Dickin, A. P., S. Moorbath and H. J. Welke. 1981. Isotope, trace element and major element geochemistry of Teriary igneous rocks, Isle of Arran, Scotland. *Trans.roy.Soc.Edinburgh,* **72,** 159–170.

Donner, J. J. 1957. Geology and vegetation of the late-Glacial retreat stages in Scotland. *Trans.roy.Soc.Edinburgh,* **57,** 633–646.

Donner, J. J. 1959. Late-Glacial and post-Glacial raised beaches in Scotland. *Suomalainen Tiedeacademia,* **55,** 1–25.

Donner, J. J. 1963. The late- and post-glacial raised beaches in Scotland II. *Annal.Acad.sci.Fenn.* **AIII 68,** 1–13.

Evans, A. L., F. J. Fitch and J. A. Miller. 1973. Potassium-argon age determinations on some British Tertiary rocks. *J.geol.Soc. London,* **129,** 419–443.

Evans, J. W. and C. J. Stubblefield. 1929. *Handbook of the Geology of Great Britain.* London.

Fleet, H. 1938. Erosion surfaces in the Grampian Highlands of Scotland. *Rapp.Comm.cart.surf.d'Appl.Tert.,* 91–94, *Union Geogr.Internat.,* Paris.

Flett, W. R. 1942. The contact between the granites of North Arran. *Trans.geol.Soc.Glasgow,* **20,** 180–204.

Friend. P. F., W. B. Harland and H. Gilbert Smith. 1970. Reddening and fissuring associated with the Caledonian unconformity in north-west Arran. *Proc.Geol.Assoc.,* **81,** 75–85.

Friend, P. F., W. B. Harland and J. D. Hudson. 1964. The Old Red Sandstone and the Highland Boundary in Arran. *Trans.Edinburgh geol.Soc.,* **19,** 363–425.

Geikie, A. 1897. *The Ancient Volcanoes of Great Britain.* London.
George, T. N. 1955. Drainage in the Southern Uplands: Clyde, Nith, Annan. *Trans.geol.Soc.Glasgow,* **22,** 1–34.
George, T. N. 1958. The Geology and Morphology of the Glasgow District. In "The Glasgow Region", 17–61, *British Assoc. Handbook.* Glasgow.
George, T. N. 1960. The stratigraphical evolution of the Midland Valley. *Trans.geol.Soc.Glasgow,* **24,** 32–107.
George, T. N. 1966. Geomorphic evolution in Hebridean Scotland. *Scott.J.Geol.,* **2,** 1–34.
Gibb, F. G. F. and C. M. B. Henderson. 1978a. Possible high-pressure relics within titaniferous augites in a basic sill. *Geol.Mag.,* **115,** 55–62.
Gibb, F. G. F. and C. M. B. Henderson. 1978b. The petrology of the Dippin Sill, Isle of Arran. *Scott.J.Geol.,* **14,** 1–27.
Goodwin, H. 1961. Radiocarbon dating and Quaternary history in Britain. *Proc.roy.Soc.* (B.), **153,** 287–320.
Gregory, J. W. 1913. *The Nature and Origin of Fiords.* London.
Gregory, J. W. 1915. The Permian and Triassic rocks of Arran. *Trans.geol.Soc.Glasgow,* **15,** 174–187.
Gregory, J. W. 1920. The pre-Glacial valleys of Arran and Snowdon. *Geol.Mag.,* **57,** 148–164.
Gregory, J. W. 1927. The fiords of the Hebrides. *Geogr.J.,* **69,** 192–212.
Gregory, J. W. and G. W. Tyrrell. 1924. Excursion to Arran. *Proc.geol.Assoc.,***35,** 401–423.
Gunn, W. 1897. Notes on the geology of the Isle of Arran. *Trans.Edinburgh geol.Soc.,* **7,** 268–276.
Gunn, W. 1899. On the old volcanic rocks of the island of Arran. *Trans.geol.Soc.Glasgow,* **11,** 174–191.
Gunn, W. 1903. The Geology of North Arran, South Bute and the Cumbraes. (Explanation of Sheet 21.) *Mem.Geol.Surv.G.B.*

Harker, A. 1903. Petrography of the Tertiary igneous rocks. *In* Gunn 1903, 103–127.

Harker, A. 1929. Igneous rocks of Tertiary age. *In* Evans and Stubblefield, 440–456.

Harland, W. B. and J. L. F. Hacker. 1966. "Fossil" lightning strikes 250 million years ago. *Advmnt Sci.*, **22,** 663–671.

Heddle, M. F. 1901. *The Mineralogy of Scotland* (edited by J. G. Goodchild). Edinburgh.

Henderson, C. M. B. and F. G. F. Gibb. 1977. Formation of analcime in the Dippin sill, Isle of Arran. *Miner.Mag.*, **41,** 534–537.

Herriot, A. 1972. A note on the composite intrusions of Bennan, Arran. *Proc.geol.Soc.Glasgow,* Sessions 111 and 112, 3–7.

Herriot, A. 1975. Observations on the Tighvein "Complex", Arran. *Proc.geol.Soc.Glasgow,* for 1972–1974, 7–11.

Herriot, A. 1981. The composite intrusions of Bennan, Arran: a further note. *Proc.geol.Soc.Glasgow,* for 1980–1981.

Herriot, A. 1982. The Pre-Tertiary dykes of Catacol, Arran. *Trans.Leeds geol.Assoc.*, **9,** 85–94.

Himus, G. W. 1954. *A Dictionary of Geology* (Penguin Reference Books R8). London.

Hodgsen, B. D. 1981. Magnetostratigraphy of the Tertiary igneous rocks of Arran, Scotland. *Unpubl. Ph.D. Thesis, Liverpool University.*

Holgate, N. 1978. A composite tholeiite dyke at Imachar, Isle of Arran: its petrogenesis and associated pyrometamorphism. *Miner.Mag.*, **42,** 141–142 & M1–4.

Holmes, A. 1960. A revised geological time-scale. *Trans.Edinburgh geol.Soc.*, **17,** 183–216.

Howie, R. A. and J. N. Walsh. 1981. Riebeckitic arfvedsonite and aenigmatite from the Ailsa Craig microgranite. *Scott.J.Geol.*, **17,** 123–128.

Hutton, J. 1975. *Theory of the Earth, with proofs and illustrations.* Edinburgh.

Hutton, J. 1899. *Theory of the Earth, with proofs and illustrations.* Vol. 3, edited by Sir A. Geikie and published by the Geol.Soc. of London.

Jehu, T. J. 1912. Discovery of fossils in the Chert and Black Shale Series at Aberfoyle. *Nature,* London, **89,** 347.

Jehu, T. J. and R. Campbell. 1917. The Highland Border rocks of the Aberfoyle district. *Trans.roy.Soc.Edinburgh,* **52,** 175–212.

Johnson, T. I. and I. G. Meighan. 1975. The Northern Granite Complex, Isle of Arran. *J.geol.Soc.London,* **131,** 331.

Judd, J. W. 1893. On composite dykes in Arran. *Quart.J.geol.Soc. London,* **49,** 536–564.

Kennedy, W. Q. 1958. The tectonic evolution of the Midland Valley of Scotland. *Trans.geol.Soc.Glasgow,* **23,** 107–133.

Khan, M. A. 1960. The remanent magnetization of the basic Tertiary igneous rocks of Skye. *Geophys.J.,* **3,** 45–62.

King, B. C. 1954. Gabbro tuff in the central volcanic complex of Arran. *Trans.geol.Soc.Glasgow,* **21,** 440–446.

King, B. C. 1955. The Ard Bheinn area of the central igneous complex of Arran. *Quart.J.geol.Soc.London,* **110,** 323–354.

King, B. C. 1959. Age of the granites of the Ard Bheinn area, Arran. *Proc.geol.Soc.London,* **1958–1959,** 134.

Kynaston, H. 1903. Notes on the petrography of the Old Red Sandstone igneous rocks. *In* Gunn 1903, 171–174.

Leitch, D. 1942. The Upper Carboniferous rocks of Arran. *Trans.geol.Soc.Glasgow,* **20,** 141–154.

Linton, D. L. 1949. Watershed breaching by ice in Scotland. *Trans.Inst.Brit.Geogr.,* **15,** 1–16.

Linton, D. L. 1951. Problems of Scottish scenery. *Scott.geogr. Mag.,* **67,** 65–85.

Longman, C. D., B. J. Bluck and O. Van Breeman. 1979. Ordovician conglomerates and the evolution of the Midland Valley. *Nature,* London, **280,** 578–581.

Lovell, J. P. B. 1971. Petrography and correlation of sandstones in the New Red Sandstone (Permo-Triassic) of Arran. *Scott. J.Geol.,* **7,** 162–169.

Lovell, J. P. B. 1981. Intertidal sediments in the Auchenhew Beds (Triassic) of Arran. Lett. to Eds. *Scott.J.Geol.,* **17,** 223.

Macculloch, J. 1819. *A. Description of the Western Isles of Scotland, including the Isle of Man.* London. (Arran, vol. ii, 311–439.)

Macgregor, M. 1944. In "Barytes in Central Scotland". (Wartime Pamphlet No. 38.) *Mem.Geol.Surv.G.B.* (Arran, 22–33.)

Macgregor, M. and A. G. MacGregor. 1948. The Midland Valley of Scotland (2nd edition). *Brit.reg.Geol., Inst.Geol.Sci.*

Mackinder, H. J. 1907. *Britain and British Seas.* London.

McCallien, W. J. 1937. *Scottish Gem Stones.* Glasgow. (Reprinted 1971, University Microfilms Ltd.)

McLean, A. C. and C. E. Deegan (eds.). 1978. The solid geology of the Clyde sheet (55 N/6 W). *Rep.Inst.Geol.Sci.* No. 78/9.

Meighan, I. G. 1979. The acid igneous rocks of the British Tertiary Province. *Bull.Geol.Surv.G.B.,* **70,** 10–22.

Miller, J. A. and P. E. Brown. 1965. Potassium-argon age studies in Scotland. *Geol.Mag.,* **102,** 106–134 (esp. Table 6, p. 125).

Miller, J. A. and W. B. Harland. 1963. Ages of some Tertiary intrusive rocks in Arran. *Miner.Mag.,* **33,** 521–523.

Mort, F. 1911. The sculpture of the Goatfell mountain group. *Scott.geogr.Mag.,* **27,** 623–643.

Mort, F. 1914. The sculpture of North Arran. *Scott.geogr.Mag.,* **30,** 393–404.

Mussett, A. E. 1980. British Tertiary Igneous Province probably not associated with East Greenland lavas. *Nature,* London, **284,** 376–377.

Mykura, W. 1960. In "Summary of Progress for 1959", pp. 38–39. *Mem.Geol.Surv.G.B.*

Mykura, W. 1965. The age of the lower part of the New Red Sandstone of south-west Scotland. *Scott.J.Geol.,* **1,** 9–18.

Peach, B. N. and J. Horne. 1910. The Scottish Lakes in relation to the geological features of the country. *In* Sir T. Murray and L. Pullar, *Bathymetrical Survey of the Scottish Fresh Water Lochs,* vol. 1, 439–513.

Peach, B. N. and J. Horne. 1930. *Chapters on the geology of Scotland.* London.

Piper, D. J. W. 1970. Eolian sediments in the basal New Red Sandstone of Arran. *Scott.J.Geol.,* **6,** 295–308.

Pollard, J. E. and J. P. B. Lovell. 1976. Trace fossils from the Permo-Triassic of Arran. *Scott.J.Geol.,* **12,** 209–225.

Pollard, J. E. and R. J. Steel. 1981. Intertidal sediments in the Auchenhew Beds (Triassic) of Arran, a reply. *Scott.J.Geol.,* **17,** 225.

Price, R. J. 1973. *Glacial and fluvioglacial landforms.* Oliver and Boyd, Edinburgh.

Price, R. J. 1982. The magnitude and frequency of late Quaternary environmental changes in Scotland: implications for human occupation. *Scott.archaeol.Review,* **1,** 61–72.

Pringle, J. 1940. The discovery of Cambrian trilobites in the Highland Border rocks near Callander, Perthshire. *Advanc. Sci.,* London, **1,** 252.

Pringle, J. 1942. On the relationship of the Green Conglomerate to the Margie Grits in the North Esk, near Edzell; and on the probable age of the Margie Limestone. *Trans.geol.Soc.Glasgow,* **20,** 136–140.

Ramsay, A. C. 1841. *Geology of the Island of Arran from original survey.* Glasgow.

Rao, M. S. 1958. Composite and multiple intrusions of the Lamlash–Whiting Bay region, Arran. *Geol.Mag.,* **95,** 265–280.

Rao. M. S. 1959. Minor acid intrusions and dykes of the Lamlash–Whiting Bay region, Arran. *Geol.Mag.,* **96,** 237–246.

Richey, J. E. 1961. The Tertiary Volcanic Districts (3rd Edition). *Brit.reg.Geol.* (Arran, 103–113).

Rogers, N. W. and I. L. Gibson. 1977. The petrology and the geochemistry of the Creag Dubh composite sill, Whiting Bay, Arran, Scotland. *Geol.Mag.,* **114,** 1–8.

Scott, A. 1915. A note on the pitchstone-porphyry of Glen Shurig. *Trans.geol.Soc.Glasgow,* **15,** 147–150.

Scott, A. 1918. Notes on the physiography of Arran. *Scott.geogr. Mag.*, **34,** 90–99.
Shackleton, R. M. 1958. Downward-facing structures of the Highland Border. *Quart.J.geol.Soc.London,* **113,** 361–389. (Arran, 377–378.)
Simpson, J. B. 1933. The late-Glacial re-advance moraines of the Highland Border, west of the River Tay. *Trans.roy.Soc.Edinburgh,* **57,** 633–646.
Simpson, J. B. 1951. The age of Tertiary vulcanicity in Scotland. *Brit.Assoc.Progr.,* **63.**
Sissons, J. B. 1963. Scottish raised shoreline heights with particular references to the Forth Valley. *Geogr.Annlr,* **45,** 180–185.
Sissons, J. B. 1966. Relative sea-level changes between 10,300 and 8,300 B.P. in part of the Carse of Stirling. *Trans.Inst. Brit.Geogr.,* **39,** 19–30.
Sissons, J. B. 1967. *The evolution of Scotland's scenery.* Edinburgh.
Smith, J. 1896. A new view of the Arran granite mountains. *Trans.geol.Soc.Glasgow,* **10,** 216–256.
Sutherland, D. S. (ed.) 1982. *Igneous rocks of the British Isles.* Wiley.

Tarr, R. S. 1908. Glacial erosion of the Scottish Highlands. *Scott.geogr.Mag.,* **24,** 575–587.
Teal, J. J. H. 1888. *British Petrography,* London.
Thomson, J. 1898. On the stratified rocks of the shore-line from Clauchland Point to the Cock of Arran. *Trans.geol.Soc.Glasgow,* **11,** 12–31.
Thomson, J. and E. A. Wünsch. 1877. Note on discovery of derived Carboniferous fossils in the conglomerates at the north end of Arran. *Trans.geol.Soc.Glasgow,* **5,** 313.
Thomson, R. N. 1982. Magmatism of the British Tertiary Volcanic Province (The Carnegie Review Article). *Scott.J.Geol.,* **18,** 49–107.

Tilley, C. E. 1957. A note on the pitchstones of Arran. *Geol.Mag.,* **94,** 329–333.

Tomkeieff, S. I. 1946. Magmatic Rolls. *Nature,* London, **158,** 420.

Tomkeieff, S. I. 1953. "Hutton's Unconformity", Isle of Arran. *Geol.Mag.,* **90,** 404–408.

Tomkeieff, S. I. 1961. Isle of Arran. No. 32 in Excursion Guide Series, *Geol.Assoc.*

Tomkeieff, S. I. and Margaret Longstaff. 1961. The magmatic complex at Kingscross Point, Isle of Arran. *Trans.Edinburgh geol.Soc.,* **18,** 194–201.

Tyrrell, G. W. 1915. On the boundary between the upper and lower divisions of the Red Rock series, near Whiting Bay and in the Sliddery Water, Arran. *Trans.geol.Soc.Glasgow,* **15,** 188–199.

Tyrrell, G. W. 1928. The Geology of Arran. *Mem.Geol. Surv.G.B.*

Tyrrell, G. W. 1948. *In* Arran and south-west Scotland. *Int.Geol. Congr.* Guide to Excursion A 15, 2–11.

Warrington, G. 1973. Miospores of Triassic age and organic-walled microplankton from the Auchenhew Beds, south-east Arran. *Scott.J.Geol.,* **9,** 109–116.

Watson, R. B. 1864. On the great drift beds with shells in the south of Arran. *Trans.roy.Soc.Edinburgh,* **23,** 523–546.

Whitelaw, C. E. 1910. The Castles of Arran. *In* Balfour and others, 241–251.

Whitten, D. G. A. and J. R. V. Brooks. 1972. *The Penguin Dictionary of Geology.* Harmondsworth.

Wright, W. B. 1937. *The Quaternary Ice Age.* London.

INDEX